How to . . .

get the most from your
COLES NOTES

Key Point

Basic concepts in point form.

Close Up

Additional hints, notes, tips or background information.

Watch Out!

Areas where problems frequently occur.

Quick Tip

Concise ideas to help you learn what you need to know.

Remember This!

Essential material for mastery of the topic.

COLES NOTES

How to get an *A* in . . .

Senior Physics

Work, energy & power

Sound & light

Electricity & relativity

Sample problems &

full solutions

COLES NOTES have been an indispensable aid to students on five continents since 1948.

COLES NOTES now offer titles on a wide range of general interest topics as well as traditional academic subject areas and individual literary works. All COLES NOTES are written by experts in their fields and reviewed for accuracy by independent authorities and the Coles Editorial Board.

COLES NOTES provide clear, concise explanations of their subject areas. Proper use of COLES NOTES will result in a broader understanding of the topic being studied. For academic subjects, Coles Notes are an invaluable aid for study, review and exam preparation. For literary works, COLES NOTES provide interesting interpretations and evaluations which supplement the text but are not intended as a substitute for reading the text itself. Use of the NOTES will serve not only to clarify the material being studied, but should enhance the reader's enjoyment of the topic.

© Copyright 1998 and Published by
COLES PUBLISHING. A division of Prospero Books
Toronto - Canada
Printed in Canada

Cataloguing in Publication Data
Henderson, Brian, 1943-

How to get an A in—senior physics

ISBN 0-7740-0564-5

1. Mathematical physics—Problems, exercises, etc.
I. Title. II. Series.

QC20.82.H46 1998   530.15'076   C98-930453-1

Publisher: Nigel Berrisford
Editing: Paul Kropp Communications
Book design and layout: Karen Petherick, Markham, Ontario

Manufactured by Webcom Limited
Cover finish: Webcom's Exclusive DURACOAT

Contents

Senior physics
Introduction

Physics is all around us. Physics is about why raindrops fall down, why airplanes and satellites stay up, why rainbows are colored and why the night sky is dark. When you were a small child, you were interested in physics. Do you remember all those "why" questions you used to ask? Many of those questions were about physics: *Why is fire hot? Why is the sky blue? Why is water wet?* Physics asks those "why" questions and provides some answers.

Though physics has a reputation as a difficult subject, anyone with curiosity, a desire to learn and a few basic math skills can enjoy and be successful in the subject. But first you have to learn the language of physics. You would not expect to be good at German if you didn't know some German vocabulary. Likewise, in physics you must learn the vocabulary of physics, and that vocabulary is mathematics.

The **Golden Rule** of mathematics is this: **Whatever is done to one side of an equation must also be done to the other side.** This is vitally important in the manipulation of physics equations.

Example 1:

To solve for v_1 in the equation $v_2 = v_1 + at$ subtract **at** from both sides.

$$v_2 - \mathbf{at} = v_1 + at - \mathbf{at}$$
$$\text{Thus: } v_1 = v_2 - at$$

Note: This is the same as transposing the **at** across the equal sign. If you are not confident of your mathematical ability, use the **Golden Rule**. As your confidence increases, switch to the faster **transposition** method.

Example 2:

To solve for t in the equation $d = \dfrac{(v_1 + v_2)}{2} t$ multiply **both** sides by **2**.

$$2 \times d = \dfrac{(v_1 + v_2)}{2} t \times 2$$

Now divide **both** sides by $(\mathbf{v_1} + \mathbf{v_2})$: $\dfrac{2d}{(v_1 + v_2)} = \dfrac{(v_1 + v_2)}{(v_1 + v_2)} t$

Thus: $t = \dfrac{2d}{(v_1 + v_2)}$

Again, the same result can be had by transposition, but using the Golden Rule produces fewer mistakes.

SIGNIFICANT FIGURES

There are several different rules regarding significant figures. For the sake of simplicity, we will adopt the following conventions:

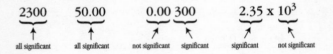

2300	50.00	0.00 300	2.35 x 10³
↑	↑	↗ ↖	↗ ↖
all significant	all significant	not significant significant	significant not significant

ADDITION AND SUBTRACTION:

> The result (answer) when numbers are added (or subtracted) should have the **same number of decimals** as the **least precise** number in the calculation.

2458.023
  340.2 ← least precise number (one decimal place)
+   2.56
2800.783

Proper answer = 2800.8 (one decimal place)

MULTIPLICATION AND DIVISION:

> The result (answer) when numbers are multiplied (or divided) should have the **same number of significant figures** as the **least accurate** number in the calculation.
>
> 1258.548
> × 45.2 ← least accurate number (3 significant figures)
> 56886.3696 Proper answer = 5.69×10^4 (3 significant figures)

CONVERSION OF UNITS

Unit conversion has always been a troublesome task for science students. By using **unit factors**, unit conversion can be greatly simplified. A unit factor is a ratio made from a conversion that is accurately known. Two examples of conversions that are accurately known are (1000 mm = 1.0 m) and (60 s = 1.0 min). Each of these simple conversions creates two unit factors.

Conversions	Unit Factors
1000 mm = 1.0 m	$\left[\dfrac{1000 \text{ mm}}{1.0 \text{ m}}\right]$ and $\left[\dfrac{1.0 \text{ m}}{1000 \text{ mm}}\right]$
60 s = 1.0 min	$\left[\dfrac{60 \text{ s}}{1.0 \text{ min}}\right]$ and $\left[\dfrac{1.0 \text{ min}}{60 \text{ s}}\right]$

To convert units, simply choose the unit factor that will cancel the unwanted unit. Multiply the unit to be converted by this factor. Continue until all the unwanted units have been replaced. Note that multiplying by a unit factor is equivalent to multiplying by one. The numerator of the factor is the same value as the denominator.

Example 1:
Convert 12 m/min to mm/s.

Solution:

$$12\,\frac{\text{m}}{\text{min}} \left[\frac{1000 \text{ mm}}{1.0 \text{ m}}\right]\left[\frac{1.0 \text{ min}}{60 \text{ s}}\right]$$

unit factors

$$= \frac{12 \times 1000}{60}\,\frac{\text{mm}}{\text{s}}$$

$$= 200\,\frac{\text{mm}}{\text{s}}$$

Example 2:

Convert 2.5 mm/s² to m/min².

Solution: $2.5 \dfrac{mm}{s^2} \left[\dfrac{1.0 \text{ m}}{1000 \text{ mm}}\right]\left[\dfrac{60 \text{ s}}{1.0 \text{ min}}\right]\left[\dfrac{60 \text{ s}}{1.0 \text{ min}}\right]$

$$\underbrace{\qquad\qquad\qquad\qquad\qquad}_{\text{unit factors}}$$

$$= \dfrac{2.5 \times 60 \times 60}{1000} \dfrac{m}{min^2}$$

$$= 9.0 \dfrac{m}{min^2}$$

These conversions may seem relatively simple. In the future, however, you will find that unit factors are very powerful tools that can be used to solve conversions that are much more challenging.

PROBLEM SOLVING

Many beginning physics students are frustrated by their lack of success in solving physics problems. Fortunately, your success in solving problems *can* improve. For those who know the technique, problem solving can become interesting and fun. Here are some ways to improve your problem-solving ability:

- Learn the definitions of the terms used in the course. You cannot be a good problem solver if you don't understand the question.
- Read the problem carefully, phrase by phrase. If there is something you don't understand, check your notes and text for clarification. Ask your friends for assistance. See your teacher for help.
- Draw a diagram to illustrate the problem. (Keep the diagrams simple. Use blocks for cars, stick people, etc.)
- On the diagram, list the information (quantities) given. Use the same symbols for the quantities that are used by your teacher and your text.
- On the diagram, list quantities that you are asked to calculate.
- Make a list of any formulas you think might be useful in the solution of the problem. Look for a connection between the given information and the formulas listed.
- Try to fit the problem into a category (e.g., uniform motion, Newton's Second Law, conservation of energy, etc.).

- Look for problems in your notes and text that are similar to the one you are trying to solve.
- List any facts or quantities you think are needed to solve the problem but are not given.

Don't expect all problems to be easy. If all problems were easy, problem solving wouldn't be interesting. Try to think of problems as interesting puzzles and challenges rather than dull, boring work. A proper attitude can go a long way to improving your chances of success. Most importantly, don't give up. With effort comes success.

PRACTICE EXERCISE:

1. Write the following numbers correct to two significant figures.
   a) 0.003587   b) 2456   c) 12.38×10^6
   d) 20054   e) 1.463×10^{-3}

2. Convert
   a) the speed 50 km/h to m/s
   b) the acceleration 10 m/s^2 to km/min^2

3. Do the calculations below and write the answer to the correct number of significant figures.

   a) 284.732   b) 1068.2148   c) 487.6   d) <u>5.874</u>
      <u>+25.3   </u>   <u>−56   </u>   <u>× 0.0037</u>   2.5

   e) $(3.784 \times 10^3)(4.1 \times 10^{-2}) =$

Mechanics

Mechanics is the study of motion. The study of motion is divided into two parts: kinematics and dynamics. Kinematics is the study of the different ways that things move. Dynamics is the study of the forces that produce the different types of motion. An important concept in the study of physics is the difference between a scalar and a vector quantity.

A **scalar** is any quantity that has magnitude (a number) and a unit (a measure), e.g., 4 metres, 5 newtons, 2 dozen, 20°C.

A **vector** is any quantity that has magnitude, a unit and a direction, e.g., 2 metres [up], 10 newtons [left].

Distance (d), a scalar quantity, is the length of the path travelled.

Displacement ($\vec{d}$) a vector quantity, is the straight line separation from the starting point to the end point.

Example:

On a walk to a friend's house, you might follow the curved path shown below.

distance d = 12 km   **displacement** $\vec{d}$ = 6 km [E]

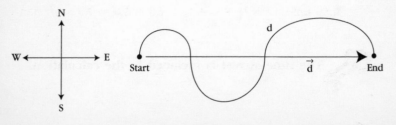

Speed (v), a scalar quantity, is the distance travelled per unit of time.

$$\text{speed} = \frac{\text{distance}}{\text{time}} \qquad v = \frac{d}{t}$$

Velocity ($\vec{v}$), a vector quantity, is the displacement moved per unit of time.

$$\text{velocity} = \frac{\text{displacement}}{\text{time}} \qquad \vec{v} = \frac{\vec{d}}{t}$$

Example:

If the trip to your friend's house (illustrated above) took 3.0 hours, find:

a) the speed for the trip   b) the velocity for the trip

Solution:

a) speed = $\frac{\text{distance}}{\text{time}} = \frac{d}{t} = \frac{12}{3} = 4$ km/h

b) velocity = $\frac{\text{displacement}}{\text{time}} = \frac{\vec{d}}{t} = \frac{6}{3} = 2$ km/h [E]

Average speed is the **total** distance travelled per unit of time.

Average velocity is the **total** displacement moved per unit of time.

Example:

A girl walks 200 m [E] in 4.0 minutes. She stops for 5.0 minutes. She then walks 500 m [W] in 9.0 minutes.

What is her: a) average speed?   b) average velocity?

Solution:

a) Total distance = $d_1 + d_2 + d_3$ = 200 + **0** + 500 = 700 m
   Total time   = $t_1 + t_2 + t_3$ = 4.0 + **5.0** + 9.0 = 18 min.

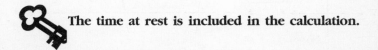

The time at rest is included in the calculation.

$$Vav = \frac{d \text{ total}}{t \text{ total}} = \frac{700}{18} = 38.8 = 39 \text{ m/minute}$$

b) Total displacement $= \vec{d}_1 + \vec{d}_2 + \vec{d}_3$

$$= 200 \text{ m [E]} + 0 + 500\text{m [W]}$$

$$= 300 \text{ m [W]}$$

$$\vec{V}av = \frac{\vec{d}\text{total}}{t \text{ total}} = \frac{300}{18} = 16.6 = 17 \text{ m/minute [W]}$$

SLOPE OF A LINE

Slope is an important concept in mathematics and physics. The slope of a graph is the steepness of the graph. Slope measures the amount of change that occurs in the dependent variable (the quantity plotted on the vertical axis) for a given change in the independent variable (the quantity plotted on the horizontal axis).

On a graph of y vs. x:

$$\text{Slope} = \frac{\text{Rise}}{\text{Run}}$$

$$= \frac{\Delta y}{\Delta x}$$

$$= \frac{y_2 - y_1}{x_2 - x_1}$$

On a graph of displacement vs. time:

$$\text{Slope} = \frac{\text{Rise}}{\text{Run}}$$

$$= \frac{\Delta \vec{d}}{\Delta t}$$

$$= \frac{\vec{d}_2 - \vec{d}_1}{t_2 - t_1}$$

$$= \text{Velocity}$$

On a displacement vs. time graph
slope = velocity.

On a graph of velocity vs. time:

$$\text{Slope} = \frac{\text{Rise}}{\text{Run}}$$

$$= \frac{\Delta \vec{v}}{\Delta t}$$

$$= \frac{\vec{v}_2 - \vec{v}_1}{t_2 - t_1}$$

$$= \text{Acceleration}$$

8

On a velocity vs. time graph
slope = acceleration.

UNIFORM MOTION

Constant speed occurs whenever an object moves **equal** distances in **equal** time intervals.

Uniform motion occurs whenever an object moves with **constant velocity** (constant speed and direction).
Note that objects moving with constant speed are not moving with **uniform motion** unless they are travelling in a straight line.

GRAPHS OF UNIFORM MOTION

Displacement vs. time graphs:

Objects moving with **uniform motion** (a hockey puck sliding across perfectly smooth frictionless ice) have constant velocity. Velocity is the slope of the displacement vs. time graph. Thus the slope of the displacement vs. time graph is a constant. Graphs with constant slope are straight lines.

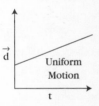

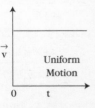

Any straight line on a displacement vs. time graph is **proof** of **uniform motion**.

Velocity vs. time graphs:

For uniform motion, **velocity is constant**. Thus the velocity vs. time graph is a horizontal straight line.

Acceleration vs. time graphs:

For uniform motion, **velocity is constant**. There is no acceleration. Thus the acceleration vs. time graph shows a = 0.

ACCELERATED MOTION:

Acceleration occurs whenever an object changes its (velocity) **speed** and/or **direction**.

Uniform acceleration occurs whenever an object makes **equal** velocity changes in **equal** time intervals.

GRAPHS OF UNIFORM ACCELERATION
Displacement vs. time graphs:

An object that is accelerating (imagine a rocket taking off) is **changing** its velocity. Velocity is the slope of the displacement vs. time graph. Thus, for an object that is accelerating uniformly, the slope of the displacement vs. time graph must be changing. Displacement vs. time graphs for accelerating objects must be curves.

The slope of this graph is increasing. On a displacement vs. time graph, **slope = velocity**. Thus this object is speeding up (accelerating)

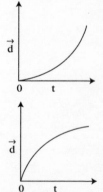

The slope of this graph is decreasing. This object is slowing down (decelerating).

VELOCITY VS. TIME GRAPHS:

Remember, acceleration is the slope of the velocity vs. time graph. Objects accelerating uniformly have constant acceleration. Thus, for objects moving with **uniform acceleration**, the slope of the velocity vs. time graph must be constant. Velocity vs. time graphs of **uniform acceleration** must be straight (non-horizontal) lines.

This object is speeding up with uniform acceleration.

slope = $\vec{a}$ = constant

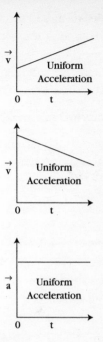

This object is slowing down uniformly (decelerating). This is also an example of uniform acceleration.

ACCELERATION VS. TIME GRAPHS:

For uniform acceleration, the acceleration is constant. Thus the acceleration vs. time graph of uniform acceleration must be a horizontal straight line.

EQUATIONS OF UNIFORM ACCELERATION:

In the following development:

$\vec{v_1}$ = initial velocity
$\vec{v_2}$ = final velocity
$\vec{a}$ = acceleration
$\vec{d}$ = displacement
t = time

Equation 1   The first equation for uniform acceleration is developed from the $\vec{v}$ vs. t graph.

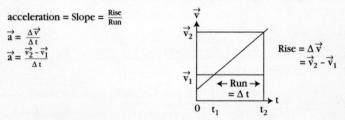

acceleration = Slope = $\frac{\text{Rise}}{\text{Run}}$

$\vec{a} = \dfrac{\Delta \vec{v}}{\Delta t}$

$\vec{a} = \dfrac{\vec{v_2} - \vec{v_1}}{\Delta t}$

Rise = $\Delta \vec{v}$
$= \vec{v_2} - \vec{v_1}$

Multiplying both sides by Δt and then adding $\vec{v_1}$ to both sides gives equation #1

#1 $\boxed{\vec{v_2} = \vec{v_1} + \vec{a}t}$ Note: Δt and t are equivalent.

11

Equation 2 The second equation for uniform acceleration is also developed from the $\vec{v}$ vs. t graph.

$$\text{displacement} = \text{Area} = \mathbf{A_1 + A_2}$$

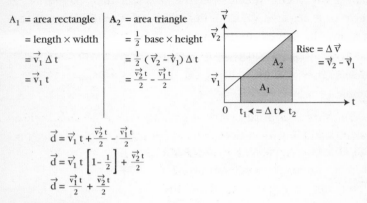

A_1 = area rectangle

= length × width

= $\vec{v}_1 \, \Delta t$

= $\vec{v}_1 \, t$

A_2 = area triangle

= $\frac{1}{2}$ base × height

= $\frac{1}{2} (\vec{v}_2 - \vec{v}_1) \, \Delta t$

= $\frac{\vec{v}_2 t}{2} - \frac{\vec{v}_1 t}{2}$

$\vec{d} = \vec{v}_1 t + \frac{\vec{v}_2 t}{2} - \frac{\vec{v}_1 t}{2}$

$\vec{d} = \vec{v}_1 t \left[1 - \frac{1}{2} \right] + \frac{\vec{v}_2 t}{2}$

$\vec{d} = \frac{\vec{v}_1 t}{2} + \frac{\vec{v}_2 t}{2}$

Factoring gives equation #2

#2 $\boxed{\vec{d} = \left[\frac{\vec{v}_1 + \vec{v}_2}{2} \right] t}$

Equation 3 Can be developed by substituting #1 into #2.

#3 $\boxed{\vec{d} = \vec{v}_1 t + \frac{1}{2} \vec{a} t^2}$

Equation 4 Can be developed by solving #1 for $\vec{v}_1$ and substituting the result into #2.

#4 $\boxed{\vec{d} = \vec{v}_2 t - \frac{1}{2} \vec{a} t^2}$

Equation 5 Can be developed by solving #1 for t and substituting the result into #2.

#5 $\boxed{\vec{v}_2{}^2 = \vec{v}_1{}^2 + 2 \vec{a} \vec{d}}$

These five equations describe **uniform acceleration**. They are all **vector** equations. Vectors have **two** features: **magnitude** (the number) and **direction**. Direction information *must* be substituted into these equations in order for them to work properly.

Direction information is put into equations by use of a **sign convention**. A sign convention is simply a statement in your problem solution about which direction is positive and which is negative. Any vector quantity in the problem then has either a positive or a negative value — depending on the direction of the vector and the agreed upon sign convention. When using the five uniform acceleration equations to solve problems, a sign convention must always be included.

Example:

The driver of a car which is moving east at 25 m/s applies the brakes and begins to decelerate at 2.0 m/s^2 (accelerate west at 2.0 m/s^2). How far does the car travel in 8.0 s?

Solution:

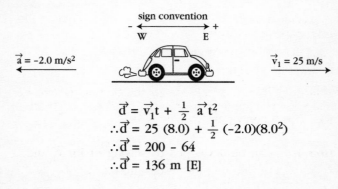

$$\vec{d} = \vec{v_1}t + \frac{1}{2}\,\vec{a}\,t^2$$
$$\therefore \vec{d} = 25\,(8.0) + \frac{1}{2}\,(-2.0)(8.0^2)$$
$$\therefore \vec{d} = 200 - 64$$
$$\therefore \vec{d} = 136\ \text{m}\ [\text{E}]$$

MOTION GRAPHS (GENERAL CONSIDERATIONS)

Displacement vs. Time Graphs:

By definition, average velocity is the ratio of total displacement to total time. On

$$\vec{V}_{av} = \frac{\vec{d}\,\text{Total}}{t\,\text{Total}} = \frac{\Delta\vec{d}}{\Delta t} = \frac{\text{Rise}}{\text{Run}} = \text{Slope}$$

a straight line $\vec{d}$ vs. t graph, (uniform motion), this calculation is the slope of the straight line. In a more general type of motion, where the $\vec{d}$ vs. t graph is a curve, the above calculation is the slope of the **secant** to the curve. (A **secant** is any straight line that cuts a curve at two or more points.)

When the interval of time becomes very small, the average velocity becomes the instantaneous velocity (the velocity an object has at a given instant). As the interval of time becomes very small, the secant becomes a **tangent**.

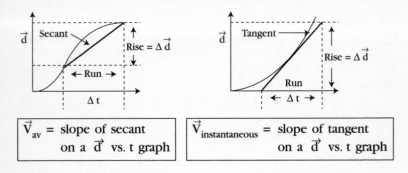

$\vec{V}_{av}$ = slope of secant on a $\vec{d}$ vs. t graph

$\vec{V}_{instantaneous}$ = slope of tangent on a $\vec{d}$ vs. t graph

Velocity vs. Time Graphs:

By definition, average acceleration is the ratio of velocity change to time

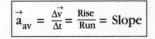

$$\vec{a}_{av} = \frac{\Delta \vec{v}}{\Delta t} = \frac{Rise}{Run} = Slope$$

required. On a straight line $\vec{v}$ vs. t graph, (uniform acceleration), this calculation is the slope of the straight line. In a more general type of motion, where the $\vec{v}$ vs. t graph is a curve, the above calculation is the slope of the secant to the curve.

When the interval of time becomes very small, the average acceleration becomes the instantaneous acceleration (the acceleration an object has at a given instant). As the interval of time becomes very small, the secant becomes a **tangent**.

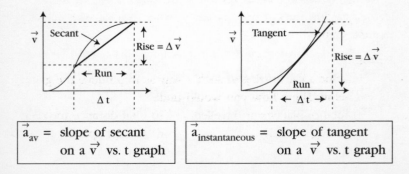

$\vec{a}_{av}$ = slope of secant on a $\vec{v}$ vs. t graph

$\vec{a}_{instantaneous}$ = slope of tangent on a $\vec{v}$ vs. t graph

VELOCITY VS. TIME GRAPHS AND AREAS

Look at what happens when we multiply velocity × time. Velocity has units of metres/second and time has units of seconds.

velocity × time = $\frac{\text{metres}}{\text{second}}$ × second

Thus: velocity × time = metres

The product of velocity and time produces a quantity that has units: metres. The quantity that has the unit "metres" is **displacement**. On velocity vs. time graphs, the product of velocity × time can be thought of as the **area** under (between the time axis and the graph) the velocity vs. time graph. We are using the term "area" not in its traditional sense of meaning "surface area," but in a more general sense that area is the product of two dimensions (in this case velocity and time) which are measured perpendicular to one another.

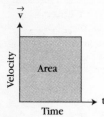

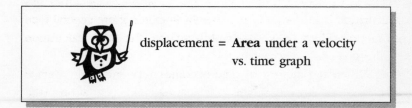

displacement = **Area** under a velocity vs. time graph

This concept, that displacement can be determined by doing an area calculation, can be very useful when dealing with velocity vs. time graphs.

Example:

Imagine a car moving in such a way as to produce the graph shown below. Show how you would find:

a) total displacement travelled
b) total distance travelled
c) average velocity
d) average speed

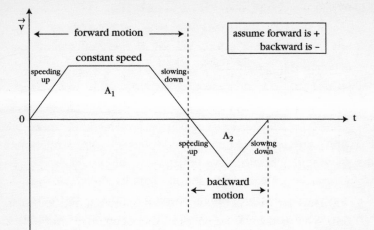

Solution:

a) Total displacement = forward displacement - backward displacement

$$= A_1 - A_2$$

b) Total distance = distance travelled forward + distance travelled backward

$$= A_1 + A_2$$

c) Average speed = $\dfrac{\text{total distance travelled}}{\text{total time}}$ $\qquad V_{av} = \dfrac{A_1 + A_2}{t}$

d) Average velocity = $\dfrac{\text{total displacement}}{\text{total time}}$ $\qquad \vec{V}_{av} = \dfrac{A_1 - A_2}{t}$

PRACTICE EXERCISE

1. A girl rides her bicycle from home to her friend's house. Starting from rest, she accelerates uniformly for 10 s. She then travels a constant velocity for 30 s. She then applies her brake and decelerates uniformly to a stop in 5.0 s. Draw a sketch (no calculations required) of her:

   a) displacement vs. time graph

   b) velocity vs. time graph

   c) acceleration vs. time graph

2. The velocity vs. time graph shown below is for a railway locomotive moving back and forth along a straight track.

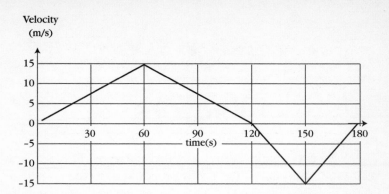

a) How far (distance) does the locomotive move in 180 s?
b) What is the displacement of the locomotive after 180 s?
c) What is the average speed of the locomotive?
d) Calculate the locomotive's average velocity.

3. A car starts out from rest and accelerates uniformly at the rate of 4.0 m/s^2. How long does the car take to travel 100 m?

4. A stone when dropped (v_1 = 0 m/s) accelerates downward at the rate of 10m/s^2. What is the velocity of the stone after is has fallen 50 m?

5. An airplane accelerating down a runway produces the displacement vs. time graph shown below. What is the instantaneous velocity of the airplane at t = 8.0 s?

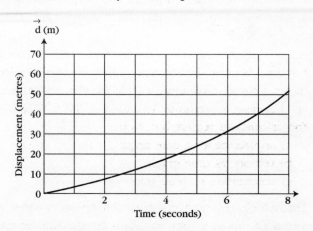

CHAPTER TWO

Vectors

Vector quantities can be represented by directed line segments. The length of the line is proportional to the magnitude of the vector quantity. The direction of the line indicates the direction of the vector quantity. **Equivalent vectors** have the **same length** and the **same direction.**

$$\vec{A} = \vec{B} \qquad \vec{A} \qquad \vec{B}$$

The inverse of a vector $\vec{C}$ is $-\vec{C}$. **Inverse vectors** have the **same length** but **opposite directions**.

$$\vec{C} \qquad -\vec{C}$$

Vector problems can be solved in two different ways. The simplest way to solve most vector problems is to draw a scale diagram and measure the required vector lengths and directions. You will need a ruler and a protractor if you use this method. The second way to solve vector problems is to use trigonometry. You should choose the method that best suits your skills and mathematical abilities.

ADDITION OF VECTORS

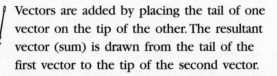

Vectors are added by placing the tail of one vector on the tip of the other. The resultant vector (sum) is drawn from the tail of the first vector to the tip of the second vector.

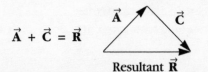

$$\vec{A} + \vec{C} = \vec{R}$$

Resultant $\vec{R}$

If more than two vectors are added, simply place the tail of each vector in sequence on the tip of the last. Draw the resultant (sum) from the tail of the first vector to the tip of the last. Numbers can be added in any order: $2 + 3 = 3 + 2$. The same is true of vectors. Vectors can be added in any order. The sum (resultant vector) is the same regardless of the order in which the vectors are added.

$$\vec{A} + \vec{C} = \vec{C} + \vec{A}$$

SUBTRACTION OF VECTORS

$$\vec{A} - \vec{C} = \vec{A} + (-\vec{C})$$

Examination of the above equation reveals that subtraction of vectors is the equivalent of addition of the inverse vector.

Example:

$$\vec{A} - \vec{C} = \vec{A} + (-\vec{C}) = \vec{D}$$

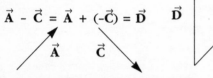

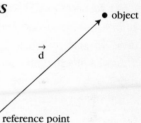

DISPLACEMENT (POSITION) VECTORS

Any object can be located by specifying the object's distance and direction with respect to a reference point. Distance and direction determine what is usually called a **displacement vector**.

VELOCITY VECTORS

The instantaneous velocity of an object is the velocity of an object at a particular moment in time. While driving in a car, combining the reading on the speedometer with your direction of travel would give the instantaneous velocity. **Instantaneous velocity vectors** are always **tangents** to the path of the object.

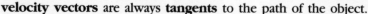

ACCELERATION VECTORS

Acceleration occurs whenever an object changes its velocity. When an object that is moving forward experiences a forward accelerating force (e.g., a car accelerating away from a stop light), the object will speed up (accelerate forward). When an object that is moving forward experiences a backward accelerating force (e.g., a car braking for a stop light), the object will slow down (decelerate).

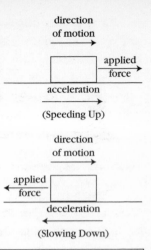

direction of motion

applied force

acceleration

(Speeding Up)

direction of motion

applied force

deceleration

(Slowing Down)

If an object's $\vec{v}$ and $\vec{a}$ vectors point in the **same** direction, the object is speeding up (accel-erating). If an object's $\vec{v}$ and $\vec{a}$ vectors point in **opposite** directions, the object is slowing down (decelerating).

RESULTANT VECTORS

Often, two (or more) vectors act on an object at the same time. The net effect is the **vector sum** (resultant) of the vectors. This can involve displacement vectors, or velocity vectors. All of these problems can be solved by applying a very simple equation.

For displacement vector problems: $\quad \vec{d_R} = \vec{d_1} + \vec{d_2}$

For velocity vector problems: $\quad \vec{v_R} = \vec{v_1} + \vec{v_2}$

Example (involving displacement vectors):

A girl rides a bicycle 10 km [N]. She then rides 20 km [W]. What is her resultant displacement?

Solution:

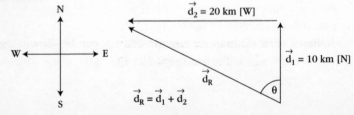

$\vec{d_2} = 20$ km [W]

$\vec{d_1} = 10$ km [N]

$\vec{d_R}$

$\vec{d_R} = \vec{d_1} + \vec{d_2}$

Using a scale diagram or trigonometry, it can be shown that:
$$\vec{d}_R = 22 \text{ km [N 63° W]}$$

Careful: When solving for a vector quantity, the answer **must** include both a **magnitude** and a **direction**.

Objects can also be affected by two or more velocity vectors. When an airplane flies through the air, the airplane has a velocity due to the push of its engines, but the air is also moving due to the wind. The resultant velocity of the aircraft relative to the ground is the vector sum of these two velocity vectors.

A person swimming across a river is affected by two velocity vectors: the velocity the swimmer acquires due to swimming action and the velocity the swimmer has due to motion of the water (the current) in the river.

Example (involving velocity vectors):

An aircraft flies with a heading (direction of the nose of the airplane) of north 30° east and an airspeed of 180 km/h. The wind is blowing west at 60 km/h. What is the resultant velocity of the aircraft? (Resultant velocity is the velocity of the plane relative to the ground.)

Solution:

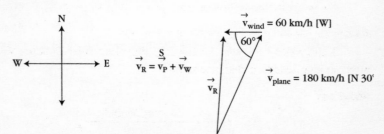

Using a scale diagram or trigonometry, it can be shown that:
$$\vec{v}_R = 159 \text{ km/h [N 11° E]}$$

PRACTICE EXERCISE

1. A rocket is moving through the upper atmosphere at 600 m/s on a path inclined at 60° to the horizontal. What is the rocket's:
   a) horizontal component of velocity?
   b) vertical component of velocity?

2. In a bicycle repair shop, a wheel of radius 32 cm is spun so that is makes one revolution in 5.0 s. A small spider is clinging to the rim of the wheel as it spins. What is the:
   a) speed of the spider?
   b) velocity of the spider when the spider is at the
           i) top? ii) side? iii) bottom?

3. A dog walks 2.0 km north, 3.0 km west and then 6.0 km south. The time for the entire trip is 120 min.
   a) What is the dog's average speed?
   b) What is the dog's average velocity?

4. An airplane flies with an airspeed of 150 km/h. The nose of the airplane points north. A wind is blowing at 60 km/h toward the west. What is the resulting velocity of the airplane with respect to the ground?

5. A boy swims across a river which is 100 m wide and flows east at 1.2 m/s. The boy starts from the south shore and swims with his body pointing north. His velocity relative to the water is 2.0 m/s.
   a) What is the resultant velocity of the boy with respect to the shore?
   b) How long does the boy take to cross the river?

CHAPTER THREE

Forces

Dynamics is the study of the forces that cause various types of motion. The underlying principles of dynamics were first developed by Sir Isaac Newton (1642-1727). Newton was a genius. His revolutionary discoveries in physics, astronomy and mathematics changed the course of science.

NEWTON'S FIRST LAW

Newton's First Law is about uniform motion and balanced forces. Balanced forces are pairs of **equal** and **opposite** forces that are applied to the **same** object.

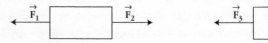

balanced forces   unbalanced forces

Newton's First Law
If the forces that act on an object are balanced, then the object **must** be in a state of uniform motion.

Any object that is at rest or moving at constant velocity is obeying Newton's First Law.

Example:

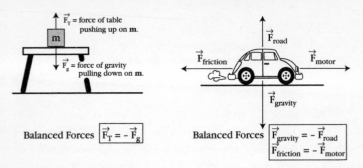

A mass at rest on a table.

$\vec{F}_T$ = force of table pushing up on **m**.

$\vec{F}_g$ = force of gravity pulling down on **m**.

Balanced Forces $\boxed{\vec{F}_T = -\vec{F}_g}$

A car moving at constant velocity.

$\vec{F}_{road}$

$\vec{F}_{friction}$

$\vec{F}_{motor}$

$\vec{F}_{gravity}$

Balanced Forces $\boxed{\begin{aligned}\vec{F}_{gravity} &= -\vec{F}_{road}\\ \vec{F}_{friction} &= -\vec{F}_{motor}\end{aligned}}$

NEWTON'S SECOND LAW

Newton's Second Law is about accelerating objects and unbalanced forces. If the forces that act on an object are not balanced, they cannot cancel one another. The force that is left over when unequal forces act on an object is the unbalanced force.

Newton's Second Law

Objects that have unbalanced forces acting on them must be accelerating. The magnitude of the acceleration varies directly as the magnitude of the applied unbalanced force. (The harder you push, the higher the acceleration.) The magnitude of the acceleration varies inversely to the mass of the object. (Pushing on a "heavy" object produces a lower acceleration than pushing on a "light" object.) The direction of the acceleration is the same as the direction of the unbalanced force.

Example:

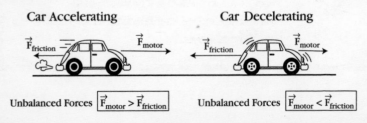

Car Accelerating

$\vec{F}_{friction}$ $\vec{F}_{motor}$

Unbalanced Forces $\boxed{\vec{F}_{motor} > \vec{F}_{friction}}$

Car Decelerating

$\vec{F}_{friction}$ $\vec{F}_{motor}$

Unbalanced Forces $\boxed{\vec{F}_{motor} < \vec{F}_{friction}}$

The mathematical version of Newton's Second Law is:

$$\text{acceleration} = \frac{\text{unbalanced force}}{\text{mass}}$$

$$\vec{a} = \frac{\vec{F}_{un}}{m}$$

Multiplying BOTH sides by **m**
produces the more common form. $\vec{F}_{un} = m\,\vec{a}$

Unit of force: $F = m\,a = kg \times \frac{m}{s^2} = \text{newton}$

Example:

A boy pushes on a 100 kg crate with a horizontal force of 500 N. The floor exerts a horizontal frictional force of 200 N on the crate. What is the acceleration of the crate?

Solution:

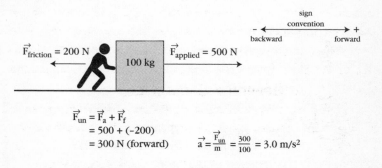

$$\vec{F}_{un} = \vec{F}_a + \vec{F}_f$$
$$= 500 + (-200)$$
$$= 300 \text{ N (forward)} \qquad \vec{a} = \frac{\vec{F}_{un}}{m} = \frac{300}{100} = 3.0 \text{ m/s}^2$$

NEWTON'S THIRD LAW

Newton's Third Law is about equal and opposite forces that are applied to **different** objects.

Newton's Third Law

For every force that acts on one object (sometimes this is called the action force), there is an **equal** and **opposite** force (sometimes called the reaction force) acting on a **different** object.

Examples:

1. Earth, Moon System:

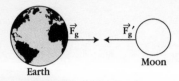

Earth   Moon

2. Firing a Gun

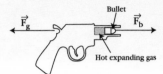

Bullet

Hot expanding gas

$\vec{F_g}$ and $\vec{F_g}'$ are EQUAL and OPPOSITE forces.

$\vec{F_g}'$ causes the moon to orbit the earth.
$\vec{F_g}$ attracts the earth toward the moon and causes the tides.

$\vec{F_B}$ and $\vec{F_G}$ are EQUAL and OPPOSITE forces.

$\vec{F_B}$ is an unbalanced force that accelerates the bullet.
$\vec{F_G}$ is an unbalanced force that accelerates the gun backwards causing the "kick" (recoil).

FORCE OF GRAVITY

The force of gravity acting on a mass is called the weight of the object. Since weight is a force, weight can be determined using Newton's Second Law.

weight = force of gravity = mass × acceleration due to gravity

$$w = m\, a_g \qquad (F = m\, a)$$

ACCELERATION DUE TO GRAVITY

From the above equation $a_g = \frac{w}{m}$. We are all familiar with the fact that if the mass of an object increases, the weight of the object is increased. If the mass of an object is tripled, then the weight will triple. In the above equation:

$$a_g = \frac{3w}{3m} \quad \text{The 3 cancels!}$$

Thus the acceleration due to gravity is the same for 3m as it is for m. This holds for any increase or decrease in mass.

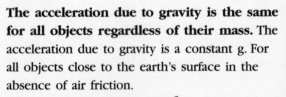

The acceleration due to gravity is the same for all objects regardless of their mass. The acceleration due to gravity is a constant g. For all objects close to the earth's surface in the absence of air friction.

$$g = 9.8 \text{ m/s}^2$$

Example:

A stone is dropped down a well which is 50 m deep. How long does the stone take to reach the bottom?

Solution:

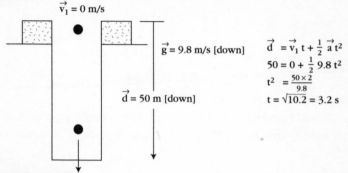

$\vec{v_1} = 0$ m/s

$\vec{g} = 9.8$ m/s [down]

$\vec{d} = 50$ m [down]

$\vec{d} = \vec{v_1} t + \frac{1}{2} \vec{a} t^2$

$50 = 0 + \frac{1}{2} 9.8 \, t^2$

$t^2 = \frac{50 \times 2}{9.8}$

$t = \sqrt{10.2} = 3.2$ s

PROJECTILES

A projectile is any object that is allowed to move freely in the earth's gravitational field. We will make two assumptions in order to simplify the projectile problem: 1. Air friction is negligible. 2. Projectiles stays close to the earth. Thus g, the acceleration due to gravity, stays constant at 9.8 m/s.

Projectile motion can be thought of as the combination of two different types of motion. Projectiles combine motion in the horizontal direction and motion in the vertical direction.

Horizontally: Since air friction is a negligible force, **there are no horizontal forces acting on projectiles**. Horizontally, projectiles must obey Newton's First Law. Projectiles must be in a state of uniform motion horizontally. Thus, horizontally:

$$\vec{d_h} = \vec{v_h} t$$

Vertically: The **only** vertical force acting is gravity. Thus gravity is an **unbalanced force**. Close to the earth, the force of gravity is **constant**. Constant unbalanced forces produce **uniform acceleration** (Newton's Second Law). Projectiles must be in a state of uniform acceleration vertically.

27

Thus, vertically:

$$\vec{v}_2 = \vec{v}_1 + \vec{g}\,t$$

$$\vec{d} = \left(\frac{\vec{v}_1 + \vec{v}_2}{2}\right) t$$

$$\vec{d} = \vec{v}_1 t + \frac{1}{2}\vec{g}\,t^2$$

$$\vec{d} = \vec{v}_2 t - \frac{1}{2}\vec{g}\,t^2$$

$$\vec{v}_2^{\,2} = \vec{v}_1^{\,2} + 2\vec{g}\,\vec{d}$$

When a projectile reaches its maximum height, the projectile stops for an instant. The vertical velocity is zero at maximum height.

Example:

1. A rock is thrown with a speed of 25 m/s at an angle of inclination of 30° to the horizontal.

a) What is the horizontal distance travelled?

b) What is the maximum height reached by the stone?

Solution:

a)

$$Sin30° = \frac{V_1}{V} \qquad \therefore V_1 = v\,sin30°$$

$$= 25\,(.500) = 12.5 \text{ m/s}$$

$$Cos30° = \frac{V_H}{V} \qquad \therefore V_H = V\cos°$$

$$= 25\,(.866) = 21.7 \text{ m/s}$$

Vertically: $\vec{V}_1 = +12.5$ m/s

$$\vec{g} = -9.8 \text{ m/s}^2$$

$$\vec{d} = 0$$

$$\vec{d} = \vec{V}_1 t + \frac{1}{2}\vec{g}\,t^2$$

$$0 = 12.5t + \frac{1}{2}(-9.8)$$

$$\therefore -4.9\,t^2 + 12.5\,t = 0 \qquad \text{Horizontally: } d_H = V_H\,t$$

$$\therefore t\,(-4.9\,t + 12.5) = 0 \qquad\qquad = 21.7 \times 2.55$$

$$\therefore t = 0 \text{ or } t = \frac{12.5}{4.9} = 2.55s \qquad\qquad = 55.3$$

$$\therefore d_H = 55m$$

Sign Convention   up +

down –

28

b) Vertically: $\qquad \therefore \vec{V}_2^2 = \vec{V}_1^2 + 2\,\vec{a}\,\vec{d}$

At maximum height;

$\vec{V}_2 = 0 \qquad\qquad \therefore 0^2 = 12.5^2 + 2\,(-9.8)\,\vec{d}$

$\vec{V}_1 = 12.5$ m/s $\qquad\quad \vec{d} = \dfrac{-156.25}{-9.8} = 15.9$

$\vec{g} = -9.8$ m/s^2

The maximum height is $\vec{d} = 16$ m [up].

2. A golfer on a tee elevated 20 m above the fairway hits a golf ball with a speed of 35 m/s at an angle of 40° inclination to the horizontal. What is the horizontal distance travelled by the ball?

Solution:

$\sin 40° = \dfrac{V_1}{V} \quad \therefore V_1 = V\sin 40°$

$\qquad\qquad = 35\,(0.643)$

$\qquad\qquad = 22.5$ m/s

$\cos 40° = \dfrac{V_H}{V} \therefore V_H = V\cos 40°$

$\qquad\qquad = 35\,(0.766)$

$\qquad\qquad = 26.8$ m/s

Vertically: $\vec{V}_1 = 22.5$ m/s

$\qquad\quad \vec{g} = -9.8$ m/s^2

$\qquad\quad \vec{d}_V = -20$ m

$\vec{d}_V = \vec{V}_1 t + \dfrac{1}{2}\,\vec{a}\,t^2$

$-20 = 22.5\,t + \dfrac{1}{2}\,(-9.8)\,t^2$

$4.9\,t^2 - 22.5\,t - 20 = 0$

$t = \dfrac{-b \pm \sqrt{b^2 - 4\,a\,c}}{2\,a}$

$\therefore t = \dfrac{22.5 \pm \sqrt{22.5^2 - 4(4.9)(-20)}}{2\,(4.9)}$

$\therefore t = \dfrac{22.5 \pm \sqrt{898.25}}{9.8}$

$t = \dfrac{22.5 \pm 30.0}{9.8}$

$t = -\dfrac{7.5}{9.8} = -0.765s$

or

$t = \dfrac{52.5}{9.8} = 5.36s$ relevant answer

$\therefore d_H = V_H\,t$

$\qquad = 26.8\,(5.36)$

$\qquad = 143.6$

$\vec{d}_H = 144$ m

FORCES AS VECTORS

Every force has a magnitude and a direction of application. Forces are vectors. If a number of forces are applied simultaneously to an object, the resultant (unbalanced) force is the sum of all the force vectors.

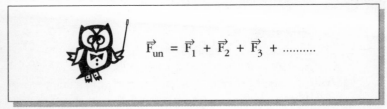

$$\vec{F}_{un} = \vec{F}_1 + \vec{F}_2 + \vec{F}_3 + ...$$

Example:

During a soccer game, three boys simultaneously kick the ball. Jeff exerts a force of 30 N [E]. Chris exerts a force of 20 N [N 30° E]. Scott exerts a force of 40 N [W]. Find the resultant (unbalanced) force acting on the ball.

Solution:

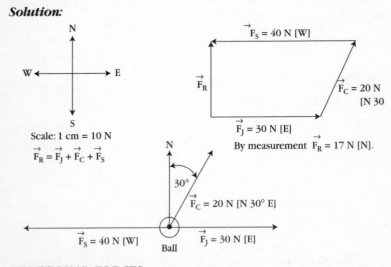

Scale: 1 cm = 10 N

$$\vec{F}_R = \vec{F}_J + \vec{F}_C + \vec{F}_S$$

By measurement $\vec{F}_R = 17$ N [N].

FRICTIONAL FORCES

Frictional forces are caused by the interaction between two surfaces in contact with one another. Frictional forces originate when one surface moves, or tends to move, relative to the other. Frictional forces almost always act opposite to the direction of motion.

Example:

A horizontal force of 20 N is applied to a mass of 5.0 kg which is at rest on a horizontal surface. The acceleration of the mass is 1.5 m/s^2. What is the force of friction on the mass?

Solution:

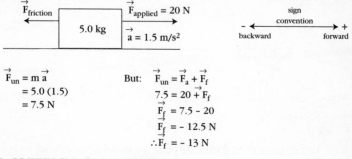

$$\vec{F}_{un} = m\,\vec{a}$$
$$= 5.0\,(1.5)$$
$$= 7.5\,\text{N}$$

But: $\vec{F}_{un} = \vec{F}_a + \vec{F}_f$
$$7.5 = 20 + \vec{F}_f$$
$$\vec{F}_f = 7.5 - 20$$
$$\vec{F}_f = -12.5\,\text{N}$$
$$\therefore \vec{F}_f = -13\,\text{N}$$

THE COEFFICIENT OF FRICTION

The coefficient of friction (μ) is the ratio of two forces, the frictional force and the normal force. The normal force (F_N) is the force a surface exerts on a mass in the direction perpendicular to the surface.

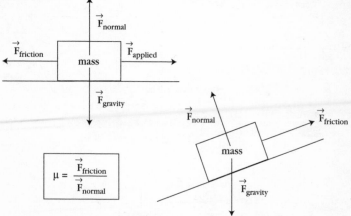

$$\mu = \frac{\vec{F}_{friction}}{\vec{F}_{normal}}$$

The coefficient of friction is a constant for a given surface.

Example:

A 10 kg mass is pushed across a wooden floor by a horizontal force of 75 N. The coefficient of friction is $\mu = 0.40$.

a) What is the force of friction?

b) What is the acceleration of the mass?

Solution:

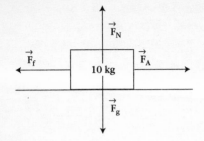

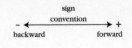

a) $F_g = mg = 10(9.8)$

$= 98N$

There is no vertical acceleration

$\therefore F_N = F_g$ (vertical forces have

equal magnitude)

$\therefore F_N = 98N$

$F_f = \mu F_N$

$= 0.40\ (98)$

$\vec{F}_f = 39.2 = 39N\ [Left]$

b) $\vec{F}_{UN} = \vec{F}_A + \vec{F}_f$

$= 75 + (-39.2)$

$= 35.8N$

$\vec{a} = \dfrac{\vec{F}_{UN}}{m} = \dfrac{35.8}{10} = 3.58$

$\vec{a} = 3.6\ m/s_2$

UNIVERSAL GRAVITATION

It was Sir Isaac Newton, in the 17th century, who first suggested that the force that caused an apple to fall from a tree is the same force that reaches out over astronomical distances and holds the moon and planets in their orbits.

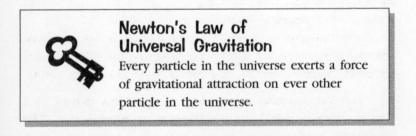

Newton's Law of Universal Gravitation

Every particle in the universe exerts a force of gravitational attraction on ever other particle in the universe.

The magnitude of the gravitational force between two objects is given by the formula shown below:

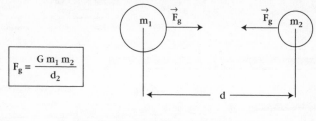

$$F_g = \frac{G\, m_1\, m_2}{d_2}$$

G = Newton's Universal Gravitation Constant

$$= 6.67 \times 10^{-11} \; \frac{N\, m^2}{kg^2}$$

Note that close to the earth the force of gravity is called weight, and is found using the formula **w = m g**. Close to the earth, the formula shown above (in the box) gives the same result as the weight calculation. As we move to greater distances from the earth only the universal gravitation equation (in the box) works because the acceleration due to gravity **g** does not remain constant.

CIRCULAR MOTION

Recall the definition of acceleration. Acceleration occurs whenever an object changes its **velocity**. Velocity is a **vector** and thus has **magnitude** (the speed of the object) and **direction** (the direction of the object). Acceleration occurs if the **speed and/or direction** of an object changes. Objects that are moving in circular paths are

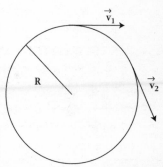

constantly changing direction. Thus, even though the speed may be constant, the objects are accelerating (because they are changing direction. Imagine a satellite in a circular orbit around the earth).

The instantaneous acceleration of an object moving in a circular path is called the **centripetal acceleration** $\vec{a}_c$. The centripetal acceleration is an instantaneous acceleration, and it always points toward the center of the circle.

$$a_c = \frac{v^2}{R} = \frac{4\pi R}{T^2}$$

a_c = acceleration
v = speed
R = radius of circle
T = period of rotation

According to Newton's Second Law, all accelerations are caused by unbalanced forces. The unbalanced force that causes the centripetal acceleration is called the **centripetal force**. The centripetal force is an unbalanced force that acts toward the center of the circle.

$$\vec{F}_c = m\,\vec{a}_c$$

ORBITAL MOTION (PLANETS AND SATELLITES)

Johannes Kepler (1571-1630) is thought by some to be the first of the "modern" astronomers. Kepler used mathematics to study and explain the motion of the planets. Kepler's studies led him to some important discoveries regarding planetary motion.

Kepler's Laws of Planetary Motion

1. The planetary orbits are ellipses. (The sun is at one focus of the ellipse.)

2. The orbital radius line sweeps out **equal** areas in **equal** intervals of time.

Time to move in orbit from a → b is the same as the time to move from c → d.
∴$A_1 = A_2$

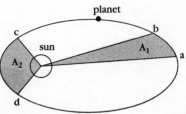

3. All bodies in orbit around the **same** central body have the **same** value of **k** where

$$k = \frac{R_o^3}{T^2}$$

R_o = orbital radius
T = period of orbit

All the planets in orbit around the sun have the same value of **k**. The **k** value for the moon is a different value because the moon is in orbit around the earth not the sun. All the satellites of Jupiter have the same value of **k** because they all orbit Jupiter.

NEWTON'S CONTRIBUTION TO PLANETARY MECHANICS

Newton realized that since bodies in orbit moved in curved paths, these orbiting objects must have centripetal forces acting on them. What, thought Newton, could be the cause of this centripetal force? There was only one possibility. The centripetal force must be the gravitational force of attraction of the sun pulling on the orbiting body. For all orbiting objects:

> **centripetal force = gravitational force of attraction**
> Fc = Fg

Example:

1.   Develop an equation for the speed of an earth satellite in an orbit of radius R_0.

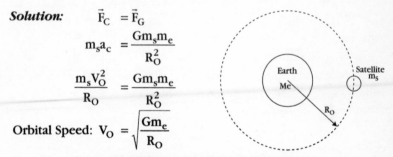

Solution:

$$\vec{F}_C = \vec{F}_G$$

$$m_s a_c = \frac{G m_s m_e}{R_O^2}$$

$$\frac{m_s V_O^2}{R_O} = \frac{G m_s m_e}{R_O^2}$$

Orbital Speed: $V_O = \sqrt{\dfrac{G m_e}{R_O}}$

2.   From basic principles, verify Kepler's Third Law.

$$F_C = F_G$$

$$\therefore m_s a_c = \frac{G m_s m_e}{R_O^2}$$

$$\therefore \frac{4\pi^2 m_s R_O}{T^2} = \frac{G m_s m_e}{R_O^2}$$

$$\therefore \frac{R_O^3}{T^2} = \frac{G m_e}{4\pi^2}$$

But all the values on the right hand side of this equation are constants.

$$\therefore \frac{R_O^3}{T^2} = K \text{ (Kepler's Third Law)}$$

WEIGHTLESSNESS

We have all seen wonderful pictures of astronauts in space as they orbit the earth in a space shuttle. The astronauts are seen floating around inside the shuttle or even outside on missions to fix broken satellites. Newscasters frequently use the term "weightless" when describing the astronauts. Unfortunately, this is incorrect. The astronauts are not weightless. Use Newton's equation for universal gravitation to determine the force of gravity acting on the astronauts in orbit 500 km above the earth. You may be surprised to find that the gravitational force on the astronauts in orbit is nearly the same as when they are standing on the earth's surface. Gravitational force reaches out over astronomical distances. Only when we move thousands of kilometres away from the earth does the gravitational force begin to decrease significantly.

How is it possible, then, for the astronauts to float around in space? The answer is gravity. Remember that objects falling in a gravitational field all fall (in the absence of air friction) with the **same** acceleration (regardless of the mass of the falling body). Near the earth's surface, the acceleration due to gravity is $g = 9.8$ m/s^2. At a height of 500 km, the acceleration due to gravity is about 8.6 m/s^2. So, even in orbit, all objects are falling at the **same** rate. Objects that always move at the **same** rate, move together. This is what is happening in the shuttle. The shuttle and everything in orbit near the shuttle (astronauts, tools and other satellites) are all falling at the same rate. The correct term for describing this is **free fall**.

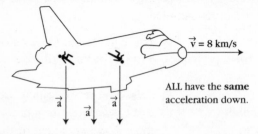

ALL have the **same** acceleration down.

In orbit, things are not weightless, they are in free fall (accelerating toward the earth at the same rate). The reason the shuttle does not fall to the earth and crash is because, as it falls, the shuttle also moves forward at very high speed (almost 8 km/s). Thus the shuttle does not fall straight down, but rather, because of its forward speed, the shuttle behaves more like a projectile and follows a curved path. The surface of the earth, of course, is also a curve. In a stable circular orbit, the curvature of the falling shuttle's path matches the curvature of the earth. So the shuttle maintains a constant separation between itself and the earth's surface.

EQUILIBRIUM

An object in equilibrium is an object moving with uniform motion. An object in equilibrium has no unbalanced forces acting on it. Therefore, the sum of **all** the forces acting on an object that is in equilibrium is **always zero**.

$$\vec{F}_{un} = \vec{F}_1 + \vec{F}_2 + \vec{F}_3 + \dots = 0$$

 Note that the above equation is a vector equation. In order to solve this equation, a vector diagram must be used. The vector diagram can be solved by drawing it to scale or applying trigonometry.

Example:

1. During a hockey game, three players simultaneously exert horizontal forces on the puck. While the forces are applied, the puck does not move. (The puck is in equilibrium.) Player one exerts a force of 30 N [W]. Player two exerts a force of 40 N [N]. What force must the third player be exerting to keep the puck motionless?

Solution:

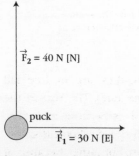

The equilibrium force equation says:

$$\vec{F}_1 + \vec{F}_2 + \vec{F}_3 = 0$$

Following this equation, we add $\vec{F}_1$ to $\vec{F}_2$ (see force diagram). The third force $\vec{F}_3$ when added to these two must produce **zero** as the sum of all three forces. Thus, the third force $\vec{F}_3$ must be drawn from point **a** to point **b**.

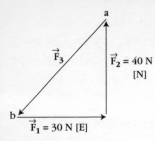

a

$\vec{F}_3$

$\vec{F}_2 = 40$ N
[N]

b

$\vec{F}_1 = 30$ N [E]

Force Diagram

$\vec{F}_3$ may now be determined by drawing the diagram to scale or using trigonometry.

Answer: $\vec{F}_3 = 50$ N [Down 37° to the Right]

Example:

2. A 50 kg sign is attached to the wall of a building by a cable and a rigid bar as shown. Find the tension in the cable.

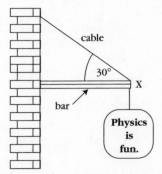

cable

30°

X

bar

Physics is fun.

Solution:

The point **x** is in equilibrium. The forces acting on **x** in this situation are: the tension $\vec{T}$ in the cable, the force of gravity $\vec{F}_g$ pulling the sign down and the force of the bar $\vec{F}_b$ pushing the sign out from the wall.

Applying the force equilibrium equation:

$$\vec{T} + \vec{F}_g + \vec{F}_b = 0$$

we get the diagram shown below.

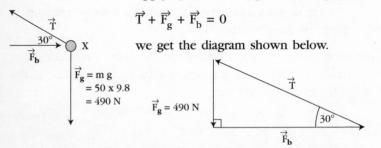

$\vec{T}$

30°

$\vec{F}_b$

X

$\vec{F}_g = m\,g$
$= 50 \times 9.8$
$= 490$ N

$\vec{F}_g = 490$ N

$\vec{T}$

30°

$\vec{F}_b$

We can now use trigonometry or redraw the diagram to scale to determine the magnitude of the vector $\vec{T}$.

Answer: T = 980 N

PRACTICE EXERCISE

1. A wagon of mass 25 kg is acted upon by a horizontal force of 80 N. What is the acceleration of the wagon, assuming friction is negligible?

2. The coefficient of friction between a 50 kg crate and a horizontal floor is $\mu = 0.46$. What is the force of friction acting on the crate?

3. A golfer hits a ball at a speed of 60 m/s from a horizontal fairway. The ball leaves the club at an angle of 30° to the horizontal. How far does the ball travel before landing on the ground?

4. Two men attach ropes to a tree stump. The men attempt to pull the stump out of the ground by pulling on the ropes. On man pulls north with a force of 4000 N. The other person pulls east with a force of 6000 N.
   a) What is the resultant force the men exert on the stump?
   b) Assuming the stump does not move, what force must the stump be exerting in order to remain in equilibrium?

5. What is the force of gravitational attraction between two cars each with a mass of 1.5×10^3 kg when the cars are parked 25 m apart?

6. Pluto is 40 times further from the sun than the earth. How long (in earth years) does Pluto take to orbit the sun once? (Hint: Use Kepler's Third Law.)

7. What is the orbital speed of the Hubble Space Telescope which is in orbit 500 km above the earth? (Assume the radius of the earth is 6400 km.)

CHAPTER FOUR

Work

Some people think reading a book, doing a math problem or giving a class presentation are examples of work. It would be very difficult, if not impossible, to measure the amount of work done in situations like these. In physics, however, we have a very precise definition of work which makes it relatively easy to quantify.

> **Work** occurs whenever a force acts to cause an object to move. (If there is no motion, there is no work done.) The amount of work done is equal to the product of the applied force and the distance moved.
>
> **work done = applied force × distance moved**
> $$W = F_A \times d$$

Metric unit of work: $W = F_A \times d$ = Newtons × metres = Joules

In cases where the applied force is not parallel to the direction of motion, we must consider the **components** of the applied force. Only the component of force that is parallel to the direction of

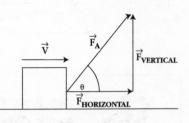

motion does work. In the case above, the vertical force does no work because there is no motion in the vertical direction. The force of gravity is strong enough to prevent the mass from being lifted off the surface by $\vec{F}_{VERTICAL}$. The **only** force doing useful work is the **horizontal** component $\vec{F}_{HORIZONTAL}$ because the motion is in the horizontal direction.

Thus, in cases where the applied force and the direction of motion are not parallel, **work done = horizontal component of force × distance moved.**

$$W = F_H \times d \qquad\qquad \text{But: } F_H = F_A \cos\theta$$

In situations where the direction of motion and the direction of the applied force are not the same, we must use the more general work equation:

$$\boxed{W = F_A \cos\theta \times d}$$

Note that when the direction of motion and the applied force are in the same direction, the general equation

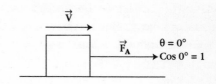

$W = F_A \cos\theta \times d$ becomes $W = F_A \times d$.

When the applied force acts perpendicular to the direction of motion,

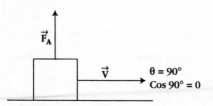

the general equation $W = F_A \cos\theta \times d$ become $W = 0$.

Any force that acts perpendicular to the direction of motion does no work.

Examples of forces that act perpendicular to the direction of motion and thus do no work are:

1. The **centripetal force** acting on objects moving along curved paths. $\vec{F}_c$ does no work.

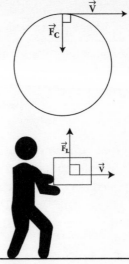

2. The **lifting force** for an object being carried horizontally across a surface. $\vec{F}_L$ does no work.

HEAT ENERGY

In situations where the force of friction acts, heat energy is created. The amount of heat energy created is equal to the

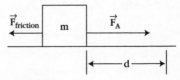

amount of work done by the force of friction.

heat energy gained = work done by friction

= force of friction × distance moved

$$\boxed{\text{heat energy } = F_f \times d}$$

ALTERNATE WORK CALCULATIONS

All of the "work" equations discussed so far depend on the forces being constant. If the applied force is not constant, other methods must be used to calculate the amount of work done. There are two other methods that can be used to calculate the amount of work done. These two alternate methods are valid for both constant and variable forces.

Alternate work calculation #1

Since work = force × distance

> work = **area** under a force vs. distance graph

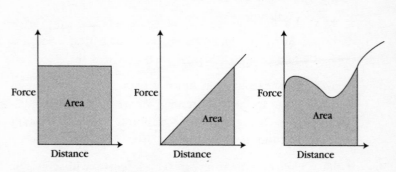

Alternate work calculation #2

work done = amount of energy gained or lost

> work = ΔE

Example:

A toy car accelerating up a ramp gains 500 J of gravitational potential energy, and 200 J of kinetic energy. Also, during the climb up the hill, 75 J of heat energy is created by frictional forces. How much work is done by the motor in the toy car?

Solution:

work done = ΔE = total energy change
$$= \Delta E_{gravitational} + \Delta E_{kinetic} + \Delta E_{heat}$$
$$= 500 + 200 + 75$$
$$= 775 \text{ J}$$

PRACTICE EXERCISE

1. A brick of mass 5.0 kg rests on a horizontal wooden floor. A boy pulls horizontally on a string attached to the brick with a force of 40 N. If the brick slides 6.0 m across the floor, how much work does the boy do?

2. If there is a force of friction of 35 N acting on the brick described in the problem above, how much heat energy is created when the brick is pulled 6.0 m across the floor?

3. A girl pulls a wagon 50 m down a path. The wagon handle she pulls on is inclined at an angle of 40° to the horizontal. If she pulls with a constant force of 100 N (directed along the length of the handle), how much work does she do?

4. If the wagon described in problem 3 moves with constant speed, what is the frictional force acting on the wagon?

5. A camel carries a 800 kg load horizontally a distance of 8.0 km across the desert. How much work does the camel do?

CHAPTER FIVE

Energy and power

Energy is a difficult term to define. Most texts define **energy** as the ability to do work. **Kinetic energy** (E_k) is the energy an object has due to its speed:

$$E_k = \frac{1}{2} mv^2$$

m = mass in kg

v = speed in m/s

Units of E_k: $E_k = kg \times \left[\dfrac{m}{s}\right]^2 = \dfrac{kg \times m \times m}{s^2} = N \times m = J$

Notice that the unit of kinetic energy is the **joule**, the same unit as work. **Potential energy** (E_p) is stored energy. There are many different forms of potential energy, e.g., the chemical energy in a battery, or the heat energy in a hot water tank.

GRAVITATIONAL POTENTIAL ENERGY

Gravitational potential energy (E_g) is the energy an object has due to its height in the earth's gravitational field. Height is usually measured from the surface of the earth. The reference height from which all heights are measured is called the **Zero of Potential Energy** (the ZPE).

$$E_g = m g h$$

m = mass in kg

g = acceleration due to gravity in m/s^2

h = height

Units of E_g: $E_g = kg \times \dfrac{m}{s^2} \times m = \dfrac{kg \times m}{s^2} \times m = N \times m = J$

The unit of gravitational potential energy (E_g) is the joule, the same unit as work and kinetic energy (E_k).

Note that this equation applies only near the earth's surface where g = constant = 9.8m/s². Far from the earth (where g ≠ constant), the calculations are more complex.

GRAVITATIONAL POTENTIAL ENERGY FAR FROM THE EARTH (where g ≠ 9.8 m/s²)

As we move further from the earth, the acceleration due to gravity and the force of gravity begin to diminish. The decrease in gravitational effects do not become significant until we reach distances of hundreds of kilometres from the earth's surface. The force of gravity decreases as illustrated by the graph shown.

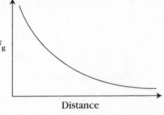

The equation for gravitational potential energy (E_g) of a mass **m** that is a great distance from the earth is:

$$E_g = - \frac{G\, m\, m_e}{d}$$

G = Newton's Universal Gravitation constant

= $6.67 \times 10^{-11} \frac{N\, m^2}{kg^2}$

m = mass of object in kg

m_e = mass of earth in kg

d = distance from earth's center

As we move to great distances from the earth, **d** becomes very large. A very large value of **d** in the above equation will cause E_g to approach the value of zero. For this reason, the ZPE (Zero of Potential Energy) when using the above equation is chosen to be at a very large distance (infinite separation) from the earth. Since everything in the universe is closer to the earth than infinite distance, everything can be thought of as being **below** the ZPE. As a result of being **below** the ZPE, everything in the universe has a **negative** value of E_g. This is one interpretation of the (–) sign in the above equation for gravitational potential energy.

Example:

How much work is done lifting (at constant speed) a 500 kg mass from a height of 1000 km above the earth to a height of 10 000 km?

(Earth radius R_e = 6400 km, Mass of earth m_e = 6.0×10^{24} kg)

Solution:

F_g is not a constant, so we cannot use $W = F \times d$ to find work done. We can, however, use $W = \Delta E = E_2 - E_1$. Since we are moving to large distances from the earth, we must use

$$E_g = -\frac{G\,m\,m_e}{d}$$

d_1 = 6 400 km + 1 000 km = 7 400 km = 7.4×10^6 m

d_2 = 6 400 km + 10 000 km = 16 400 km = 16.4×10^6 m

m = 500 kg

$d_2 = 16.4 \times 10^6$ m

$$E_2 = -\frac{G\,m\,m_e}{d_2}$$

m = 500 kg

$d_1 = 7.4 \times 10^6$ m

$$E_1 = -\frac{G\,m\,m_e}{d_1}$$

Re

Earth

$m_e = 6.0 \times 10^{-24}$ kg

$$W = \Delta E_g = E_2 - E_1$$

$$\therefore W = -\frac{G\,m\,m_e}{d_2} - \left[-\frac{G\,m\,m_e}{d_1}\right]$$

$$\therefore W = G\,m\,m_e\left[-\frac{1}{d_2} + \frac{1}{d_1}\right]$$

$$\therefore W = 6.7 \times 10^{-11} \times 500 \times 6.0 \times 10^{24}\left[-\frac{1}{16.4 \times 10^6} + \frac{1}{7.4 \times 10^6}\right]$$

$$\therefore W = 2.01 \times 10^{17}\left[7.42 \times 10^{-8}\right]$$

$$\therefore W = 1.5 \times 10^{10} \text{ joules}$$

47

CONSERVATION OF ENERGY

One of the most important principles in science is the **Law of Conservation of Energy**. This law states that energy cannot be created or destroyed, but it can be transformed from one type to another. When a rock rolls down a hill, gravitational potential energy is converted to kinetic energy (the rolling) and heat energy (because of friction). When a battery is used in a flashlight, chemical energy is converted to heat and light energy in the light bulb. In a hydroelectric generating station, water falling from an elevated position behind a dam or waterfall loses potential energy that is transformed by turbines into electrical energy. The electrical energy is distributed to individual homes and factories where it is further transformed into other types of energy: heat energy in stoves and ovens, light energy in light bulbs, mechanical energy in motors, etc.

Mathematically, the Law of Conservation of Energy can be stated as follows:

total energy input = total energy output

Example:

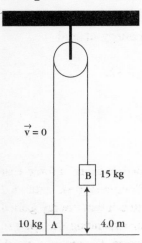

A 10 kg mass and a 15 kg mass are attached to one another by a light rope. The rope runs through a light, frictionless pulley, as shown. If the system is released from rest, what will be the speed of the masses the instant the 15 kg mass hits the ground?

Solution:

Notice that, initially, there is no kinetic energy (v = 0). Only the 15 kg mass has energy due to its height above the ground:

$$E_1 = E_{g_B} = m_B gh = 15 \times 9.8 \times 4.0 = 588 \text{ J}$$

At the instant before the 15 kg mass contacts the ground, the masses will be in the situation illustrated in diagram #2. In this situation, **both** masses are moving. Thus they both have kinetic energy. Since they are attached by a rope, they must have the same

48

speed. Also, the 10 kg mass is now elevated and thus it has gravitational potential energy. (The 15 kg mass is now at ground level and has no gravitational potential energy.)

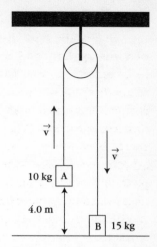

The total energy of the system just before contact with the ground is:

$$E_2 = E_kA + E_kB + Eg = \frac{1}{2} m_A v^2 + \frac{1}{2} m_B v^2 + m_A gh$$
$$= \frac{1}{2} 10v^2 + \frac{1}{2} 15v^2 + 10 (9.8) 4.0$$
$$= 12.5v^2 + 392$$

Using the Law of Conservation of Energy:

total energy at start = total energy at end
$$E_1 = E_2$$
$$588 = 12.5v^2 + 392$$
$$12.5v^2 = 196$$
$$v = \sqrt{15.7}$$
$$v = 4.0 \text{ m/s}$$

SPRINGS

As a spring is compressed or extended, it exerts a force that opposes the extension or compression. Compressed or extended springs contain stored (potential) energy. Recall that energy gained or lost ΔE = **work done** (compressing or expanding the spring). Calculating the energy stored in a spring becomes a work calculation. For most springs, we cannot use the $W = F \times d$ formula because the force exerted by a spring is not constant. If we have a graph of the force exerted by a spring (Fsp) as a function of its extension (compression) **x**, we can determine the work done (energy gained or lost) by calculating the area under the graph.

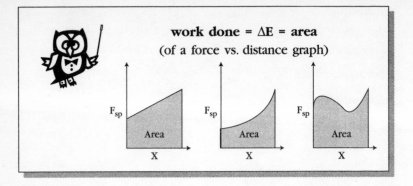

work done = ΔE = area
(of a force vs. distance graph)

There is a special class of spring called a **linear spring**. Linear springs are springs that obey Hooke's Law:

$$F_{sp} = K x$$

F_{sp} = force exerted by the spring in newtons
K = spring constant in N/m
x = spring compression or extension in metres

Springs with large values of K are very stiff like the springs in a truck suspension. Springs with small values of K are easily stretched and compressed like a Slinky.

Linear springs always produce force vs. compression graphs that are straight lines passing through the point (0,0).

Linear Spring

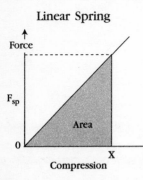

$$\text{slope} = \frac{\text{rise}}{\text{run}} = \frac{F_{sp}}{x} = K$$

(the spring constant)

The potential energy stored in the spring as it is compressed a distance **x** is:

ΔE = work done = area

E_{sp} = area of a triangle

$\quad = \frac{1}{2}$ base × height

$\quad = \frac{1}{2} (x) F_{sp} \quad$ but $F_{sp} = K x$

$E_{sp} = \frac{1}{2} (x) K x$

$$E_{sp} = \frac{1}{2} K x^2$$

Example:

1. How much work is done compressing a linear spring 0.25 m? The spring constant for the spring is 250 N/m.

Solution: work done = $E_{sp} = \frac{1}{2} K x^2 = \frac{1}{2} 250 (0.25)^2 = 7.8$ J

Example:

2. If the spring described above is the spring in a dart gun, how fast will the dart be travelling when it leaves the barrel?

Solution:

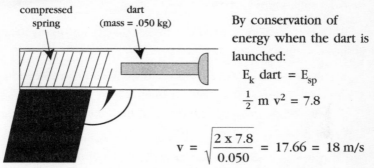

compressed spring   dart (mass = .050 kg)

By conservation of energy when the dart is launched:

$$E_k \text{ dart} = E_{sp}$$
$$\frac{1}{2} m v^2 = 7.8$$

$$v = \sqrt{\frac{2 \times 7.8}{0.050}} = 17.66 = 18 \text{ m/s}$$

Power is the rate of doing work (or using energy).

$$\text{power} = \frac{\text{work done}}{\text{time}}$$

$$\boxed{P = \frac{W}{t}}$$

unit of power; $P = \frac{\text{joules}}{\text{second}} = $ watts

A NEW ENERGY UNIT

From the above equation: $W = P \times t$

Recall that **work done = energy gained or lost**.

Thus: $\Delta E = P \times t$.

If power is measured in kilowatts and time is measured in hours, then:

$$\Delta E = P \times t = \text{kilowatts} \times \text{hours}$$

This common unit of energy is the kWh (**kilowatt hour**). We purchase energy in units of kWh.

Example:

Calculate the cost of using a 60 Watt light bulb continuously for one week. The cost of electrical energy is $.05/kWh.

Solution:

Converting 60 Watts to kilowatts:

$$60 \text{ Watts} \left[\frac{1 \text{ kilowatts}}{1000 \text{ Watts}} \right]$$

60 Watts = 0.060 kW

Converting 1 week to hours:

$$1 \text{ week} \left[\frac{7 \text{ days}}{1 \text{ week}} \right] \left[\frac{24 \text{ hours}}{1 \text{ day}} \right]$$

1 week = 168 h

$$\Delta E = P \times t$$
$$\Delta E = 0.060 \times 168 = 10.08 = 10 \text{ kWh}$$
$$\text{cost} = \text{energy used} \times \text{rate} = 10.08 \times 0.05 = \$0.504$$

It will cost 50 cents to use the light bulb continuously for a week.

PRACTICE EXERCISE

1. What is the kinetic energy of a 10 g bullet moving at 250 m/s?

2. How much potential energy is lost by each cubic metre of water that falls over Niagara Falls (a vertical drop of 60 m)? (The density of water is 1 000 kg/m^3.)

3. What is the gravitational potential energy of a 2.6×10^8 kg asteroid located 100 000 km from the earth's center? The mass of the earth is 6.0×10^{24} kg.

4. Assuming the masses are initially at rest and the pulleys and ropes are light and frictionless, with what speed will the 25 kg mass shown below hit the ground?

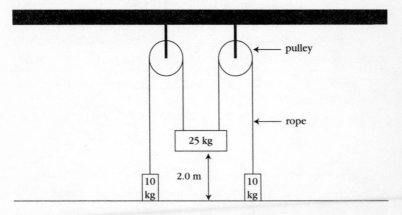

5. A 120 kg athlete training for the Olympics runs up a flight of stairs a vertical distance of 20 m in 12 seconds. What power was developed by the athlete?

6. A crossbow behaves like a linear spring with a force constant of 3500 N.
   a) How much energy is stored in the bow when the string is pulled back 0.75 m?
   b) If the bow is used to launch an arrow of mass 250 g, with what speed will the arrow leave the bow?

CHAPTER SIX

Momentum

Momentum is the product of an object's mass and velocity.

momentum = mass × velocity

$$\vec{p} = m\,\vec{v}$$ Momentum is a vector.

Units of momentum: $\vec{p} = kg \times \frac{m}{s} = \frac{kg \times m}{s}$

IMPULSE

Starting with Newton's Second Law: $\vec{F} = m\,\vec{a}$ but $\left[\vec{a} = \frac{\Delta\vec{v}}{\Delta t}\right]$

$$\therefore \vec{F} = m\,\frac{\Delta\vec{v}}{\Delta t}$$

$$\therefore \boxed{\vec{F}\Delta t = m\,\Delta\vec{v}}\quad Ⓐ$$

The left side of equation Ⓐ $\vec{F}\Delta t$ is the quantity called "impulse." **Impulse**, a vector quantity, is the measure of the strength and duration of a force. The unit of impulse is the Newton × second (Ns).

Recall that force is the product on mass × acceleration (F = ma). Thus the unit of force, the Newton, is equivalent to a kg × m/s². The impulse unit, the Ns is force × time. Thus the unit of impulse is a (kg × m/s²) × s = $\frac{kg \times m}{s}$, the same unit as momentum.

Impulse has some interesting aspects as it applies to collisions: a car hitting a tree, a hammer striking a nail, a boxer's fist impacting an opponent, etc. Consider a car of mass **m** travelling at 50 km/h colliding with a large tree and brought suddenly to rest. The mass of the car **m** and its $\Delta\vec{v}$ are fixed. Thus, for a given event, the right-hand side of equation Ⓐ is a constant. This means that $\vec{F}\Delta t$ must also be a constant value. However, there are an infinite number of possible values of $\vec{F}$ and Δt that will produce the needed product, the impulse $\vec{F}\Delta t$. A force of 2F and a time of application of $1/2\Delta t$ will result in an impulse of $\vec{F}\Delta t$. A force of 1/10F and a time of 10Δt will produce an impulse of $\vec{F}\Delta t$.

In the design of a modern automobile, for safety reasons, much consideration is given to the quantity impulse ($\vec{F}\Delta t$). Engineers design cars so that, during a collision, the time interval Δt over which the car and its passengers are brought to rest is maximized. By using shock-absorbing front ends, seat belts, air bags, padded dashboards and crumple zones, engineers have been able to increase the time interval Δt over which a car and its passengers are brought to rest. So far, these improvements have increased Δt by a factor of about four. Increasing Δt by four reduces the forces applied to the car and its passengers by the same amount [$\frac{1}{4}\vec{F} \times 4\Delta t = \vec{F}\Delta t$]. Thus the force exerted on the car passengers would be one-quarter of the force experienced by passengers in a car without the safety features mentioned above. The injuries suffered by the passengers would be greatly reduced.

The impulse equation $\vec{F}\Delta t = m\Delta\vec{v}$ can be further developed by substituting $\Delta\vec{v} = \vec{v_2} - \vec{v_1}$.

$$\therefore \vec{F}\Delta t = m\,(\vec{v_2} - \vec{v_1})$$

Equation a $\therefore$ $\boxed{\vec{F}\Delta t = m\vec{v_2} - m\vec{v_1}}$

but $\qquad \vec{p_1} = m_1\vec{v_1}$ = initial momentum

$\qquad\qquad p_2 = m_2\vec{v_2}$ = final momentum

Equation b $\therefore$ $\boxed{\vec{F}\Delta t = \vec{p_2} - \vec{p_1}}$

Equation c $\therefore$ $\boxed{\vec{F}\Delta t = \Delta\vec{p}}$

Equation d $\therefore$ $\boxed{\vec{F} = \frac{\Delta\vec{p}}{\Delta t}}$

Equations a, b and c show that impulse is related to momentum. Impulse is equal to the **change** in momentum. Equation d shows another interesting relationship. The unbalanced force acting on an object is equal to the **rate of change of momentum**.

Law of Conservation of Momentum

In all interactions between objects, the total momentum before the interaction is equal to the total momentum after the interaction.

Example:

A 5.0 kg block moving right at 2.0 m/s collides head on with a 3.0 kg block moving left at 4.0 m/s. After the collision, the 3.0 kg block is observed to be moving right at 1.0 m/s. What is the velocity of the 5.0 kg block after the collision?

In all collisions, momentum is conserved. We can use the Law of Conservation of Momentum to solve this problem. Remember, **momentum is a vector**! A sign convention or a vector diagram **must** be used.

Solution:

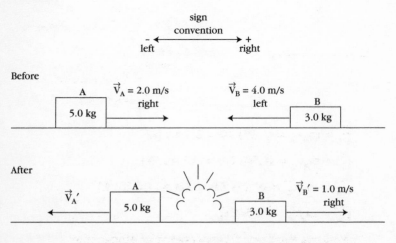

Law of Conservation of Momentum:
total momentum before = total momentum after

$$\vec{p_A} + \vec{p_B} = \vec{p_A}' + \vec{p_B}'$$

$$m_A\vec{v_A} + m_B\vec{v_B} = m_A\vec{v_A}' + m_B\vec{v_B}'$$

$$5.0\,(2.0) + 3.0\,(-4.0) = 5.0\,\vec{v_A}' + 3.0\,(1.0)$$

$$5.0\,\vec{v_A}' = -5.0$$

$$\vec{v_A}' = -1.0 \text{ m/s}$$

After the collision the 5.0 kg mass is moving left at 1.0 m/s

The problem above is an example of a **one-dimensional** collision problem. Collisions, however, often take place in two and three dimensions.

Example: (two-dimensional collision)

A 250 g ball Ⓐ moving at 10 m/s [east] collides with a 500 g ball Ⓑ moving at 6.0 m/s [north]. After the collision, the 250 g ball Ⓐ is observed to be moving at 8.0 m/s [N20°E]. What is the velocity of Ⓑ after the collision?

Solution:

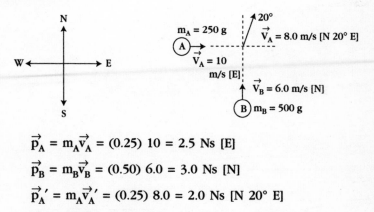

$$\vec{p_A} = m_A\vec{v_A} = (0.25)\,10 = 2.5 \text{ Ns [E]}$$

$$\vec{p_B} = m_B\vec{v_B} = (0.50)\,6.0 = 3.0 \text{ Ns [N]}$$

$$\vec{p_A}' = m_A\vec{v_A}' = (0.25)\,8.0 = 2.0 \text{ Ns [N 20° E]}$$

$$\vec{p_B}' = m_B\vec{v_B}' = (0.50)\,\vec{v_B}'$$

Applying the Law of Conservation of Momentum:
$$\vec{p_A} + \vec{p_B} = \vec{p_A}' + \vec{p_B}'$$

A vector diagram must be drawn using the above formula.

The diagram shown below is a sketch only. It is not drawn to scale.

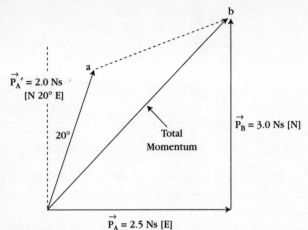

The sum of $\vec{p}_A + \vec{p}_B$ is the total momentum. $\vec{p}_A' + \vec{p}_B'$ must also equal the total momentum. Thus the missing vector $\vec{p}_B'$ must be drawn from **a** to **b**. By drawing the diagram to scale and measuring $\vec{p}_B'$, or using trigonometry we find:

$$\vec{p}_B' = 2.14 \text{ Ns [N } 58° \text{ E]}$$

$$\therefore m_B\vec{v}_B' = 2.14$$

$$\therefore 0.50 \text{ } v_B' = 2.14$$

$$\therefore \vec{v}_B' = \frac{2.14}{0.50}$$

$$\therefore \vec{v}_B' = 4.3 \text{ m/s [N } 58° \text{ E]}$$

PRACTICE EXERCISE

1. A 2.0 kg baseball initially at rest is hit by a bat. The ball moves off with a velocity of 30 m/s.
   a) What impulse does the ball experience?
   b) If the collision between the bat and the ball lasts 0.080 seconds, what force does the ball experience?

2. A car with a mass of 1.2×10^3 kg is moving east at 25 m/s. What is the momentum of the car?

3. A dart with a mass of 60 g is thrown with a speed of 12 m/s at an apple with a mass of 200 g. If the dart sticks in the apple, with what speed will the apple/dart combination move immediately after the collision?

4. A 6.0 kg bowling ball moving at 2.8 m/s east collides with a 2.0 kg bowling pin. After the collision, the pin is moving at 4.0 m/s east. What is the velocity of the bowling ball after the collision?

5. A 1.5 kg hockey puck sliding west across the ice at 6.0 m/s collides with a 2.0 kg stone which is initially at rest. After the collision the stone moves off at 3.2 m/s in the direction west 25° north. What is the velocity of the puck after the collision?

CHAPTER SEVEN

Waves

A **wave** is the transfer of energy through a carrier material called a **medium** by means of some type of vibration. There are many different types of waves. Two of the most common are the **transverse wave** and the **longitudinal wave**.

TRANSVERSE WAVES

A transverse wave is one in which the particles in the medium vibrate in a direction perpendicular to the direction of travel of the wave.

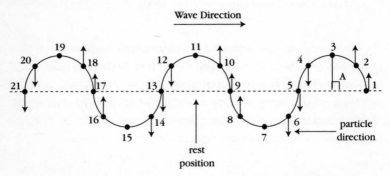

Amplitude (A) is the perpendicular distance from the rest position (the center line above) to a position of maximum displacement.

Phase is an important concept in the study of waves. Particles that are in phase have the same velocity and displacement relative to the rest position of the medium. Particles in phase are doing the same thing at the same time. Particles that are in phase in the diagram above are: (1, 9 and 17), (2, 10 and 18), (3, 11 and 19), (4, 12, and 20), etc.

Wavelength (λ) is the distance between adjacent particles that are in phase. The wavelength is the distance between particles 3 and 11, or 4 and 12, or 1 and 9, etc.

LONGITUDINAL WAVES

A longitudinal wave is a wave in which the particles in the medium vibrate parallel to the direction of motion.

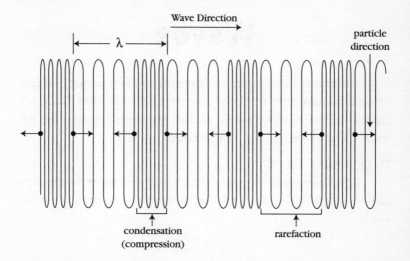

Wave Direction

particle direction

condensation (compression)

rarefaction

Sound is an example of a longitudinal wave. Seismic waves (vibrations produced by earthquakes) contain both longitudinal and transverse components. Water waves, light waves, waves in a vibrating guitar string and most other waves are transverse. All waves are vibrations. The **frequency** of vibration (f) is the number of oscillations that occur per unit of time.

$$f = \frac{\text{number of vibrations}}{\text{time taken}}$$

Unit of frequency (f) = vibrations/second
                     = cycles per second (cps)
                     = Hertz (Hz)

The **period** (T) of wave (vibration) is the time required to produce one complete wave.

$$T = \frac{1}{f} \qquad f = \frac{1}{T}$$

Unit of period (T) = seconds

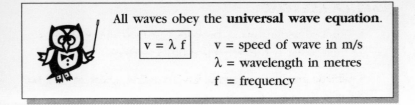

All waves obey the **universal wave equation**.

$$v = \lambda f$$

v = speed of wave in m/s
λ = wavelength in metres
f = frequency

The **principle of superposition** of waves states that when two waves occupy the same position in a medium, the particles in the medium are displaced by the algebraic sum of the displacements of the individual waves. ("Algebraic" means we must consider displacements above the rest position as being positive, while displacements below the rest position are negative in value. As a result, colliding waves can reinforce one another or cancel one another.

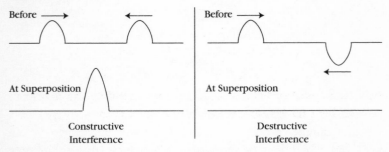

Waves behave differently depending on whether they are reflected from a **fixed** (not moveable) **end** or a **free** (easily moved) **end**.

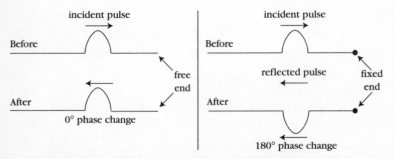

The incident pulse reflected from a free end is unchanged. This is a 0° phase change. The incident pulse reflected from a fixed end is inverted. This is a 180° phase change.

PRACTICE EXERCISE

1. A plucked violin string produces 75 000 sound waves in 1.0 min. What is the frequency of the string?

2. A whistle produces a sound having a frequency of 512 Hz. What is the period of this sound vibration?

3. Predict the shape of the reflected waves:

4. A sound wave with wavelength 0.45 m has a frequency of 740 Hz. What is the speed of the sound wave?

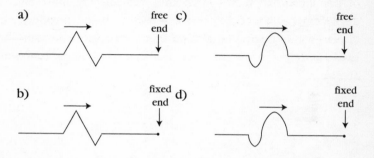

5. The speed of light is 3.0×10^8 m/s. Calculate the frequency of light having a wavelength of 6.5×10^{-7} m.

CHAPTER EIGHT

Light

Light is but a small part of a much broader phenomenon called electromagnetic radiation. All forms of radiation consisting of oscillating electric and magnetic (electromagnetic) fields are members of the electromagnetic spectrum.

Low Energy				visible light		**High Energy**	
radio	TV	micro- waves	infrared (heat radiation)	r y o g b v e e r r l i d l a e u o l n e e l o g n e w e t	ultraviolet	X-rays	gamma rays
low frequency long wavelength			←			→	high frequency short wavelength

Electromagnetic radiation (light) travels through a vacuum at very high speed. The speed of light in a vacuum is $\mathbf{c = 3.0 \times 10^8}$ **m/s**. In all other materials, light travels somewhat more slowly.

In a uniform medium such as air or water, light travels in straight lines. This straight line motion of light is called **rectilinear propagation**.

When light strikes a material, some light gets reflected, some is transmitted into the material and some light is absorbed and converted to heat. On striking a dull, dark surface, most of the light is absorbed and converted to heat. A small amount of light energy is reflected and some is transmitted into the material. On striking a transparent material like glass, most of the light is transmitted through the glass. Only small amounts are reflected and absorbed. On hitting a mirror, most of the light is reflected. Small amounts are absorbed and transmitted into the mirror material.

By passing white light through a prism, it can be shown that white light is composed of **all** of the colors of the visible light spectrum. **Dispersion** is the separation of light into its component colors by refraction.

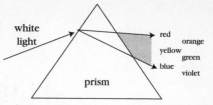

Objects appear colored because they selectively reflect or transmit some parts of the spectrum more efficiently than others. When illuminated by white light, a book that appears blue **reflects** the blue regions of the spectrum more strongly than other regions. A piece of transparent red glass **transmits** red light more efficiently than other colors.

INTENSITY OF ILLUMINATION (point light sources)

As light leaves a point source of light, it spread out in straight-lines. This straight-line spreading of light rays gives rise to an

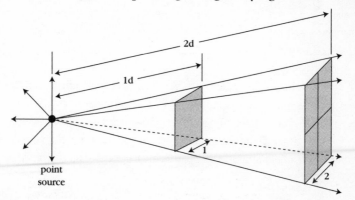

interesting relationship. It can be seen in the diagram that moving twice as far from a point light source allows the light to fall on a surface area four times bigger. We will let the amount of light energy falling on unit area at distance **d** be I. At a distance **2d**, the same amount of light energy falling one unit area at distance **d** is now falling on four (2 × 2) unit areas. Thus, at the greater distance, each unit area is receiving (1/4)I . Three times farther away, the light energy falling on each unit area would be (1/9)I. The illumination **I** is inversely proportional to the square of the distance from a point light source.

$$I \propto \frac{1}{d^2} \therefore \boxed{I = \frac{k}{d^2}}$$

where: I = illumination falling on unit area in lux (lx)

k = strength of the source of lumens (lm)

d = distance form point source in metres

Example:

A point light source (a small projector bulb) of strength **k** provides illumination of 100 lux on a screen at a distance of 2.0 m. What illumination will the same source produce at a distance of 8.0 m?

Solution:

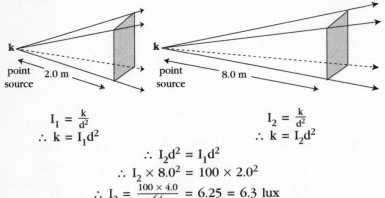

$$I_1 = \frac{k}{d^2} \qquad\qquad\qquad\qquad I_2 = \frac{k}{d^2}$$
$$\therefore k = I_1 d^2 \qquad\qquad\qquad\qquad \therefore k = I_2 d^2$$
$$\therefore I_2 d^2 = I_1 d^2$$
$$\therefore I_2 \times 8.0^2 = 100 \times 2.0^2$$
$$\therefore I_2 = \frac{100 \times 4.0}{64} = 6.25 = 6.3 \text{ lux}$$

REFLECTION

When light strikes a surface, some of the light gets reflected. The direction of the reflected light is determined by the Law of Reflection.

Law of Reflection:

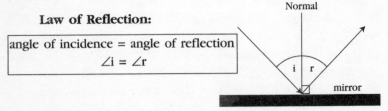

angle of incidence = angle of reflection

$$\angle i = \angle r$$

REFRACTION

When light passes from one transparent material into a different transparent material, the path of the light ray changes (refracts). The amount of refraction depends on the relative optical densities of the two materials. **Optical density** is a term which describes how a transparent material interacts with light. Materials with high optical density transmit light more slowly and refract light through larger angles than do materials with low optical density. The mathematical relationship between the original direction of the light and the final direction of the light is known as **Snell's Law**.

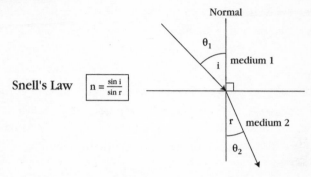

Snell's Law $\quad n = \dfrac{\sin i}{\sin r}$

The index of refraction (n) is a constant for a given pair of materials. Tables of indexes of refraction appear in most physics textbooks. The indexes of refraction that are listed in these tables usually use air or vacuum as medium 1 (the incident medium).

When light passes into a material having a higher optical density, the path of the light ray is refracted (bent) toward the normal. Light passing into a material with a lower optical density bends away from the normal.

Improved notation for Snell's Law:

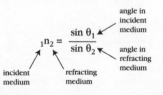

$$_1n_2 = \frac{\sin \theta_1}{\sin \theta_2}$$

For light going from air into water: $\quad _An_W = \dfrac{\sin\theta_A}{\sin\theta_W} = 1.3$

For light going from air into glass: $\quad _An_G = \dfrac{\sin\theta_A}{\sin\theta_G} = 1.5$

For light going from water into glass: $\quad _Wn_G = \dfrac{\sin\theta_W}{\sin\theta_G} = 1.2$

Example:

Light passes from water into glass at an angle of incidence of 55°. What is the angle of refraction?

Solution:

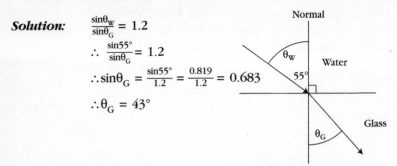

$$\frac{\sin\theta_W}{\sin\theta_G} = 1.2$$

$$\therefore \frac{\sin 55°}{\sin\theta_G} = 1.2$$

$$\therefore \sin\theta_G = \frac{\sin 55°}{1.2} = \frac{0.819}{1.2} = 0.683$$

$$\therefore \theta_G = 43°$$

When light passes from one medium into another, it follows a particular path. If the refracted ray is reversed so that it now becomes an incident ray, the light emerging into the first medium will follow the original incident ray path (but in the opposite direction).

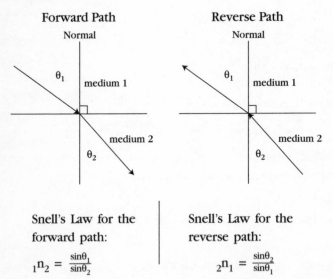

Snell's Law for the forward path:

$$_1n_2 = \frac{\sin\theta_1}{\sin\theta_2}$$

Snell's Law for the reverse path:

$$_2n_1 = \frac{\sin\theta_2}{\sin\theta_1}$$

Comparing Snell's Law for the forward and reverse paths, we see:

$$\boxed{_1n_2 = \frac{1}{_2n_1}}$$

The index of refraction is also related to the speed of light in the incident and refracting mediums.

$$_1n_2 = \frac{v_1}{v_2}$$

$_1n_2$ = index of refraction for light going from medium 1 into medium 2

v_1 = speed of light in medium 1 in m/s

v_2 = speed of light in medium 2 in m/s

Example:

Find the speed of light in water.

Solution:

We know $v_A = c = 3.0 \times 10^8$ m/s and $_An_W = 1.3$

$_An_W = \frac{v_A}{v_W}$ ∴ $1.3 = \frac{3.0 \times 10^8}{v_W}$ ∴ $v_W = \frac{3.0 \times 10^8}{1.3}$

∴ $v_W = 2.3 \times 10^8$ m/s

TOTAL INTERNAL REFLECTION

When light passes from a medium of high optical density into one with a lower optical density, the light ray bends **away** from the normal. As a result, the angle of refraction is larger than the angle of incidence. This allows for a somewhat unusual situation in

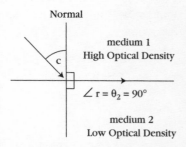

which the angle of refraction reaches its maximum possible value of 90°, while the angle of incidence is less than 90°. The angle of incidence that causes the angle of refraction to become a right angle (90°) is called the **critical angle** (C).

Example:

Find the critical angle for a water/air boundary.

Solution:

In order for there to be a critical angle, light must pass from the medium with the higher optical density (water) into the medium with the lower optical density (air). As a result we need to calculate $_Wn_A$.

$_Wn_A = \frac{1}{_An_W} = \frac{1}{1.3} = 0.77$

Applying Snell's Law at the critical angle:

$$\therefore {_w}n_A = \frac{\sin\theta_w}{\sin\theta_A}$$

$$\therefore 0.77 = \frac{\sin C}{\sin 90°}$$

$$\therefore \sin C = 0.77$$

$$\therefore C = 50°$$

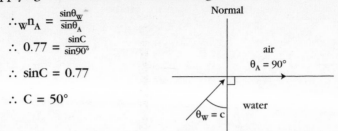

For angles of incidence greater than the critical angle, there is no refracted ray since the angle of refraction cannot be greater than 90°. All of the light energy is reflected. This is know as **total internal reflection**. Total internal reflection is responsible for confining light to the inside of a thread of glass in a fiber optics cable.

Fiber Optics Cable
High Optical Density

Low Optical Density

HOW WE SEE OBJECTS

All the objects we see are sources of light. Objects that produce their own light (a flame, the sun, a red-hot piece of iron) are luminous objects. Objects that we see because they reflect light produced by luminous objects are described as non-luminous (books, cars, trees, the moon, etc.). All the objects we see emit or reflect light in **diverging light ray cones**. The eye has the ability to focus

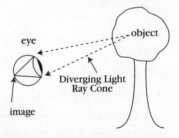

these diverging light ray cones and produce an image on the retina of the eye. Whenever a diverging light ray cone is intercepted by the eye, the eye always "sees" an image at the source of the light ray cone.

PLANE MIRRORS

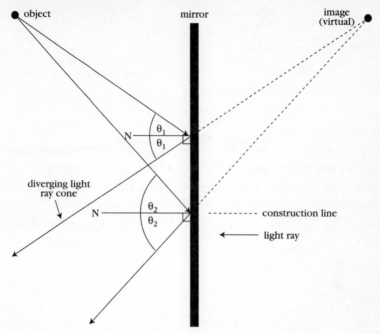

Virtual images are "apparent sources" of light ray cones. The light ray cones are formed by construction line extensions of light rays. Virtual images cannot be projected onto screens. This is why plane mirrors (like bathroom wall mirrors) cannot reflect your image onto the wall or a screen. **Real images** are formed at sources of light ray cones made from actual (real) light rays. Real images can be projected onto screens. Real images are created by movie projectors and thus the images formed can be projected onto a screen.

SPHERICAL MIRRORS

Spherical mirrors are mirrors that can be imagined to have been cut from a hollow reflecting sphere. Think of the reflections you have seen in shiny spherical Christmas ornaments. Concave spherical mirrors use the inside reflective coating. Convex spherical mirrors use the outside reflective coating.

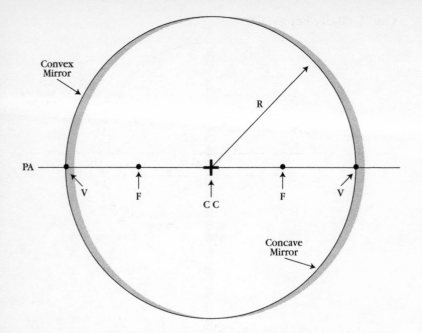

CC (Center of Curvature) The center of the reflecting sphere from which the mirror was cut.

V (Vertex) The center of the mirror.

PA (Principle Axis) The straight line drawn through the center of curvature and the vertex.

F (Focus) The midpoint between the vertex and the center of curvature.

R (Radius of Curvature) The radius of the reflecting sphere.

CONCAVE SPHERICAL MIRRORS

Rules

1. Any ray parallel to the principle axis is reflected through the focus.
2. Any ray through the focus is reflected parallel to the principle axis.
3. Any ray through the center of curvature is reflected back along the same path.

Case 1: Object beyond CC

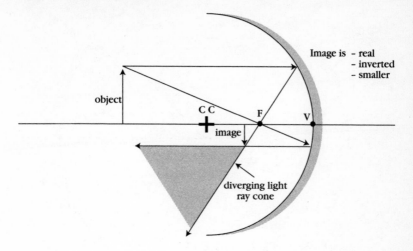

Case 2: Object between CC and F

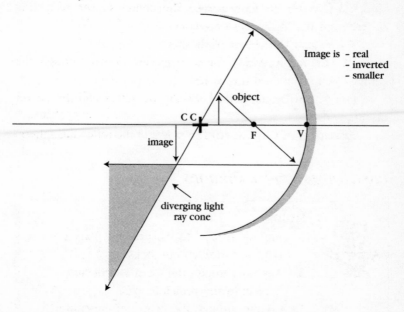

Case 3: Object between F and V

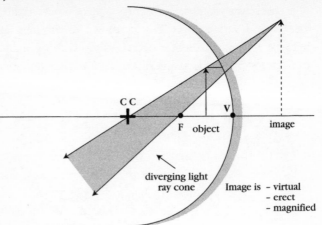

Image is - virtual
- erect
- magnified

CONVEX SPHERICAL MIRRORS

Rules

1. Any ray parallel to the principle axis is reflected away from the focus.
2. Any ray directed toward the focus is reflected parallel to the principle axis.
3. Any ray directed toward the center of curvature is reflected back along the same path.

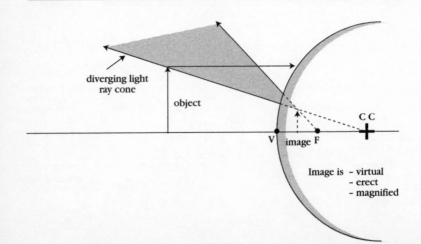

Image is - virtual
- erect
- magnified

LENSES

Lenses are formed from curved pieces of transparent material. The curved surface of a lens is able to refract light and produce light ray cones. Thus lenses are very important in image-producing optical systems such as microscopes, telescopes, binoculars, etc.

CONVEX THIN LENSES

> ## Rules
> 1. Any light ray parallel to the principle axis is refracted through the focus.
> 2. Any light ray through the focus is refracted parallel to the principle axis.
> 3. Any light ray through the vertex passes straight through.

Case 1) Object beyond F

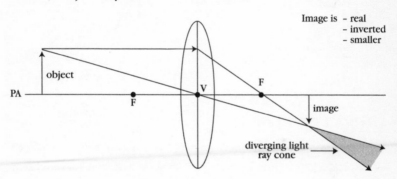

Image is   - real
                - inverted
                - smaller

Case 2) Object between F and V

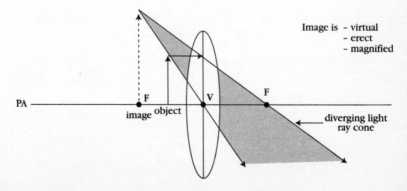

Image is   - virtual
                - erect
                - magnified

CONCAVE THIN LENSES

Rules
1. Any ray of light parallel to the principle axis is refracted away from the focus.
2. Any ray of light through the vertex passes straight through.

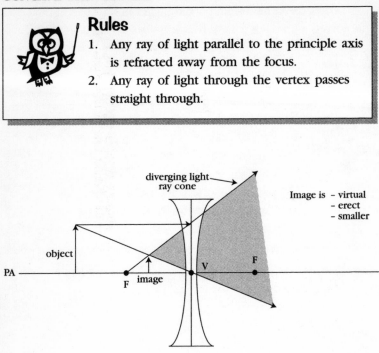

diverging light ray cone

Image is – virtual
– erect
– smaller

object

image

PA

V

F

F

PRACTICE EXERCISE

1. A projector light bulb (acting like a point light source) produces an illumination of 50 lux on a screen placed 12 m away. At what distance from the bulb must the screen be located so that the illumination increases to 100 lux?

2. Light passes from air into water at an angle of incidence of 60°. What is the angle of refraction?

3. The index of refraction for light going from air into glass ($_an_g$) is 1.5. What is the index of refraction for light passing from glass into air ($_gn_a$)?

4. What is the critical angle for a diamond? ($_an_d = 2.4$)

5. For the object reflected in the plane mirror shown below, draw the image and state its characteristics.

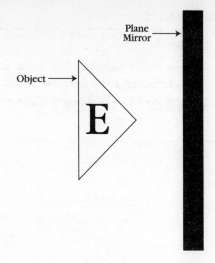

6. Use ray tracing to draw and describe the image formed by the concave mirror shown below.

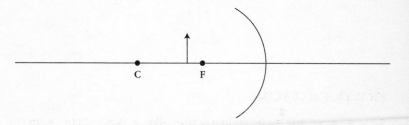

7. Use ray tracing to draw and describe the image formed by the lens shown below.

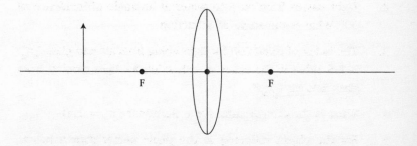

CHAPTER NINE

Interference

Interference occurs whenever two or more waves occupy the same region of space. In order for interference to occur, we need two sources of waves. In the simplest situation, the wave sources will be in phase emitting waves with the same wavelength. Imagine two small loudspeakers emitting the same musical note with the waves in phase.

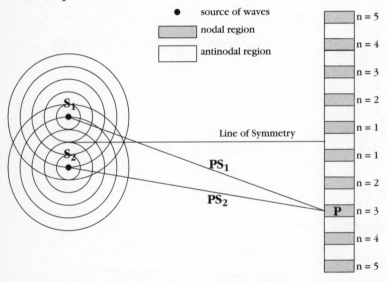

At any point **P** on the line of symmetry, since it is exactly the same distance to S_1 and S_2, wave crests from these sources will arrive at exactly the same instant. On the line of symmetry, waves will therefore reinforce one another producing a region of constructive interference. This is called an antinodal region.

As we move off the line of symmetry, the distance to one source PS_2 decreases while the distance to the other source PS_1 increases. A "path difference" (PS_1-PS_2) is produced. Because they travel different distances, wave crests from S_1 and S_2 will arrive at **P** at different times. When the path difference between the sources is $\frac{1}{2}$ λ, a wave crest from one source will arrive at **P** just when a trough is arriving from the other source. Destructive interference will occur. Waves from one source will cancel waves from the other source. A region of destructive interference is called a nodal region. This destructive interference occurs again when:

$(PS_1\text{-}PS_2)$ is $\frac{3}{2} \lambda$, $\frac{5}{2} \lambda$, $\frac{7}{2} \lambda$, etc.

In general, destructive interference occurs when:

$$\boxed{PS_1 - PS_2 = (n - \tfrac{1}{2}) \lambda}$$
PS_1 = distance from P to S_1 in m.
PS_2 = distance from P to S_2 in m.
λ = wavelength in m.
n = 1, 2, 3, 4, ...etc.
= number of nodal line.

Constructive interference occurs whenever $PS_1 - PS_2 = n \lambda$.

The equation above works well for phenomena with relatively large wavelengths, such as sound and water waves.

Example:

In an open field, two loudspeakers emit sound waves in phase. A student walks perpendicular to the line of symmetry until she reaches the 8th nodal line. At this point, she finds she is 32 m from S_1 and 28 m from S_2. What is the wavelength of the sound emitted by the speakers?

Solution:

$$PS_1 - PS_2 = (n - \tfrac{1}{2}) \lambda$$
$$\therefore 32 - 28 = (8 - \tfrac{1}{2}) \lambda$$
$$\therefore 4.0 = 7.5\lambda$$
$$\therefore \lambda = \tfrac{4.0}{7.5} = 0.53 \text{ m}$$

DOUBLE SLIT EXPERIMENT

Thomas Young (1773-1829) was the first to demonstrate interference of light waves. He managed this historic feat using a very simple apparatus. Young allowed light from a single source to fall on an opaque screen in which two pinholes had been made. The pinholes were very small, so light waves passing through the pinholes were greatly diffracted and became semicircular. The pinholes were placed very close together, so waves entered each pinhole simultaneously. The waves emerged on the other side simultaneously. As a result, each pinhole became a source of semicircular, in-phase waves superimposed on one another.

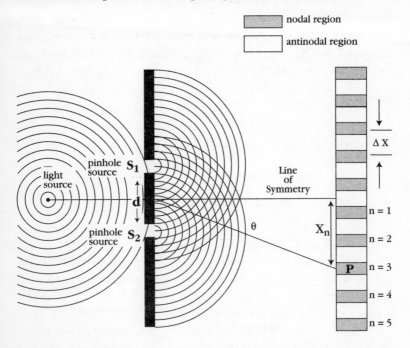

For phenomena with a short wavelength such as light, and where point **P** is far from the sources, the original equation for double source interference can be modified to

$$d \sin \theta = (n - \tfrac{1}{2}) \lambda$$ and $$\frac{dx_n}{L} = (n - \tfrac{1}{2} \lambda$$

$\lambda =$ wavelength in m
$n =$ number of nodal line
$\theta =$ angle between 'L' the line of symmetry

$d =$ distance between S_1 and S_2 in m
$x_n =$ perpendicular distance from line of symmetry to P in m
$L =$ distance from center of S_1 and S_2 to 'P' in m

The spacing between adjacent nodal lines is given by

$$\Delta x = \frac{\lambda L}{d}$$

Example:

A student uses an intense monochromatic light source to illuminate two narrow slits. The slits are 0.20 mm apart. The resulting interference pattern falls on a screen 4.0 m away from the slits. The 12th nodal line of the pattern is located 14 cm from the line of symmetry. What is the wavelength of the light used?

Solution:

$$\frac{dx_n}{L} = (n - \frac{1}{2})\,\lambda$$

$$\therefore \frac{(2.0 \times 10^{-4})(1.4 \times 10^{-1})}{4.0} = (12 - \frac{1}{2})\,\lambda$$

$$\therefore \lambda = \frac{(2.0 \times 10^{-4})(1.4 \times 10^{-1})}{11.5 \times 4.0}$$

$$\therefore \lambda = 6.09 \times 10^{-7} = 6.1 \times 10^{-7} \text{ m}$$

THIN FILM INTERFERENCE

Perhaps you have noticed the fascinating colored patterns that appear when gasoline or oil is spilt on water, or the bright iridescent colors visible is a soap bubble. Both of these effects and many others are due to the effects of interference.

We need two sources of waves to produce interference. In the case of thin film interference, the two waves' sources come from the two surfaces (front and back) of the thin film.

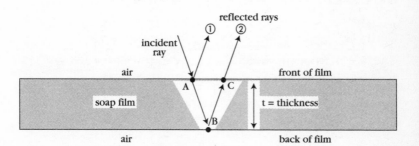

If reflected rays 1 and 2 are in phase, they will superimpose constructively and produce a bright (antinodal) reflection. If rays 1 and 2 are out of phase, they will cancel one another and produce a dark (nodal) reflection.

The **total phase difference** (TPD) between reflected rays 1 and 2 depends on two factors. The first factor is the path difference **ABC**. For light falling directly onto the film, the path difference ABC = 2 t. The second factor is "fixed end reflection." Remember, when waves are reflected from a fixed end, the reflected wave is turned upside down (a 180° phase change). Any medium that slows light down (has a high optical density, a high index of refraction) will act like a fixed end for reflected waves. At **A**, light is passing from air into (soapy) water. Water has a higher index of refraction than air. The light slows down going from air into water, so reflected ray 1 undergoes a 180° phase change. This is equivalent to a phase difference of $\frac{1}{2}\lambda$. At **B**, light is passing from water into air. The light speeds up. **B** acts like a free end for reflected rays, so there is no phase change at **B**. At **C**, the light ray is transmitted out into the air, **not** reflected. Transmitted waves never undergo phase changes. Thus we need only consider A and B when looking for fixed end reflections.

Total Phase Difference = Path Difference + $\frac{1}{2}\lambda$ for each fixed end
     1 and 2   1 and 2   (at A and B)

$$\text{TPD} = \text{ABC} + \tfrac{1}{2}\lambda \text{ for each fixed end}$$

$$\boxed{\text{TPD} = 2t + \tfrac{1}{2}\lambda \text{ for each fixed end}}$$

Example:

A thin film of oil (n_o = 1.4) covers a flat glass slide (n_g = 1.6). The oil film has a uniform thickness of 200 nm. What is the smallest wavelength of light that will be strongly reflected from the oil film?

Solution: $\boxed{\text{TPD} = 2t + \tfrac{1}{2}\lambda \text{ for each fixed end}}$

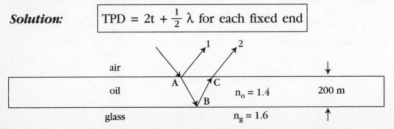

Light slows down when it passes from air into oil ($n_o > n_A$). Light slows down again when it passes from oil into glass ($n_g > n_o$).

Thus there are **two** fixed ends (one at A and one at B).
$\therefore$ TPD $= 2(2.0 \times 10^{-7}) + 2(\frac{1}{2}\lambda)$
$\therefore$ TPD $= 4.0 \times 10^{-7} + \lambda$

In order for a strong reflection to occur, TPD $= 0, \lambda, 2\lambda, 3\lambda$, etc. The smallest value of TPD that gives a solution to the equation above is TPD $= 2\lambda$.
$\therefore 2\lambda = 4.0 \times 10^{-7} + \lambda$
$\therefore \lambda = 4.0 \times 10^{-7}$

Remember, all the phase shifting occurs in the film. Thus this is the wavelength in the film.

To determine the wavelength of light in air we must use $_A n_o = \frac{v_A}{v_o}$.

But: $\quad v_A = \lambda_A f \qquad \therefore\ _A n_o = \frac{\lambda_A}{\lambda_o}$

$\qquad\qquad v_o = \lambda_o f \qquad \therefore\ \lambda_A = \lambda_o(_A n_o) = 4.0 \times 10^{-7}(1.4)$

$\qquad\qquad\qquad\qquad\qquad \therefore\ \lambda_A = 5.6 \times 10^{-7}$ m

Clearly, light is able to exhibit interference phenomena. Only waves can interfere with one another, producing regions of constructive and destructive interference. Particles like bullets, cars and rocks are not able to do this. Interference is a strong piece of evidence that light is a wave.

WAVE MODEL VS. PARTICLE MODEL OF LIGHT

Is light a stream of particles, or a series of waves? This question has intrigued scientists for hundreds of years. A brief summary of some of the evidence supporting each theory is shown in the following chart:

Key: ✓ = this property of light tends to support the indicated theory

$\qquad$ ✗ = this property of light cannot easily be explained by the indicated theory

PROPERTY	EXAMPLE	WAVES	PARTICLES	NOTES
Transmission through Vacuum	star light	✗	✓	A 19th-century physicist developed the concept of the "ether" as a medium to propegate light waves through a vacuum.
Speed	$c = 3.0 \times 10^8$ m/s	✓	✗	No particle having mass has been accelerated to speed = c.
Rectalinear Propagation	laser beam	✓	✓	The path of a light ray bends in a gravitational field.
Reflection	$\angle i = \angle r$	✓	✓	Both particles and waves obey the Law of Reflection.
Refraction	$n = \frac{\sin i}{\sin r}$	✓	✓*	Particles and waves both refract and obey Snell's Law. $\left.\begin{array}{l}\text{Waves}\\\text{Light}\end{array}\right\}$ $_1n_2 = \frac{v_1}{v_2}$ * Particles $_1n_2 = \frac{v_2}{v_1}$ $\left\{\begin{array}{l}\text{Different}\\\text{from light}\end{array}\right.$
Partial Transmission Reflection		✓	✗	Waves naturally do this, particles do not.
Color	R O Y G B V	✓	✗	Difficult to explain using particles.
Polarization		✓	✗	Difficult to explain using particles.
Dispersion		✓	✗	Waves naturally do this, particles do not.
Diffraction		✓	✗	Waves naturally do this, particles do not.
Interference		✓	✗	Waves naturally do this, particles do not.
Photo-electric Effect		✗	✓	Particles naturally do this, waves do not.

84

Today, scientists believe that light is neither wave nor particle, but has the properties of both. If we do an experiment to examine the wave nature of light, light behaves like a wave. If we try to examine the particle nature of light, light behaves like a particle. This ambidextrous property of light is sometimes referred to as light's wave-particle duality.

PRACTICE EXERCISE

1. In order to frighten crop-eating birds, a farmer places two loudspeakers in his field. The speakers are 20 m apart and they emit a steady sound consisting of waves in phase. Walking in the field the farmer notices that when he is 18 m from one speaker and 23 m form the other, he is on the 8th nodal line. What is the wavelength of the sound emitted?

2. A student doing a Young's double slit experiment passes laser light through a pair of slits 0.25 mm apart. The interference pattern is projected onto a screen 12 m away. The student observes the distance from the line of symmetry in the interference pattern to the 16th nodal line is 37 cm.
   a) What is the wavelength of the laser light?
   b) What is the spacing between the nodal lines?

3. For each of the thin films shown below, predict whether there will be a bright (antinodal) reflection or a dark (nodal) reflection.

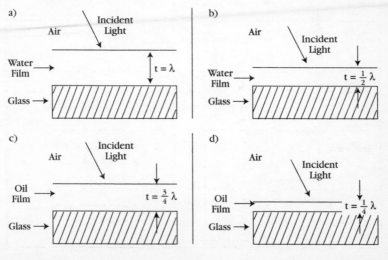

85

CHAPTER TEN

Electric charge

The world we see around us is made from matter. The structure of matter has been a subject of interest to philosophers and scientist for over three thousand years. Scientists have discovered that matter is made from atoms and atoms in turn are composed of subatomic particles called electrons, protons and neutrons. Protons and neutrons are made from yet smaller particles called quarks. For the purpose of explaining most of the basic properties of matter, it is necessary to understand some of the properties of electrons, neutrons and protons.

Particle	Symbol	Charge	Relative Mass	Actual Mass (in kg)	Location
Electron	e	−1	1	9.11×10^{-31}	in orbit around nucleus
Proton	p	+1	1800	1.67×10^{-27}	inside nucleus
Neutron	n	0	1800	1.67×10^{-27}	inside nucleus

The structure of an atom can be represented symbolically as:

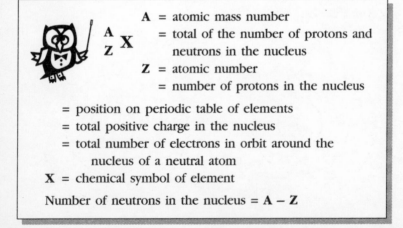

$^A_Z X$

A = atomic mass number
= total of the number of protons and neutrons in the nucleus

Z = atomic number
= number of protons in the nucleus
= position on periodic table of elements
= total positive charge in the nucleus
= total number of electrons in orbit around the nucleus of a neutral atom

X = chemical symbol of element

Number of neutrons in the nucleus = A − Z

Example:

Draw an atomic model diagram of a fluorine atom ${}^{19}_{9}\text{F}$.

Solution:

Number of neutrons = A − Z = 19 − 9 = 10

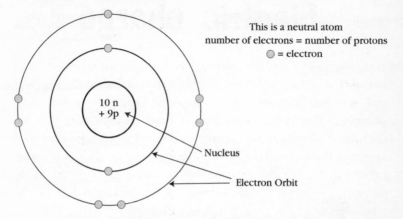

This is a neutral atom
number of electrons = number of protons
◎ = electron

10 n
+ 9p

Nucleus

Electron Orbit

All electrical phenomena must be explained in terms of the movement of electrons. A **negative charge** is caused by a surplus of electrons. A **positive charge** is created by a deficit of electrons. Protons do not move from atom to atom in electrical events. Protons are isolated in the centers of atoms, held in the nucleus by very strong nuclear forces. Electrons, on the other hand, are in orbits around the outside of the atom. Electrons in the outer orbits of some atoms can be caused to move from one atom to another by relatively modest forces. Atoms which have gained or lost electrons are called ions.

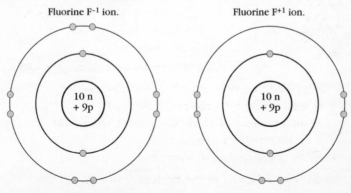

Fluorine F^{-1} ion.

Fluorine F^{+1} ion.

10 n
+ 9p

10 n
+ 9p

> The **Law of Electrostatics** states that like charges repel one another. Opposite charges attract each other.

The charge on a single electron is an extremely small amount of charge. As a result, a more convenient unit and a much larger measure for electric charge has been chosen. This larger charge unit is called the **coulomb (C)**.

The charge on one electron = 1 elementary charge
$$= 1.6 \times 10^{-19} \text{ coulombs.}$$

$$\text{Number of elementary charges} = \frac{\text{charge of coulombs}}{\text{charge of one electron}}$$

$$N = \frac{Q}{e}$$

Example:
How many electrons does it take to produce a charge of 1.0 C?

Solution:
$$N = \frac{Q}{e} = \frac{1.0}{1.6 \times 10^{-19}} = 6.25 \times 10^{18} = 6.3 \times 10^{18} \text{ elementary charges}$$

One coulomb of electric charge contains 6.3 billion, billion elementary charge.

The magnitude of the force between charged objects is given by **Coulomb's Law**.

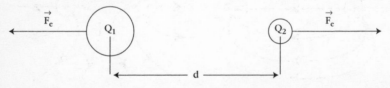

Coulomb's Law: $F_e = \dfrac{kQ_1Q_2}{d^2}$

F_e = electric force in N
Q_1 and Q_2 = charge in C
d = distance between charge centers in metres
k = Coulomb's constant
$= 9.0 \times 10^9 \ \dfrac{Nm^2}{C^2}$

(The direction of F_e is determined by applying the Law of Electrostatics. Like charges repel. Opposite charges attract.)

An **electric field** is the volume of space in which the presence of an electric charge can be detected. Electric fields are graphically represented by lines of force. Some common electric fields are shown below:

Isolated Point Charges

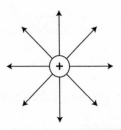

Pairs of Point Charges

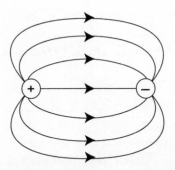
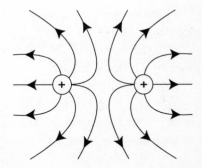

Oppositely Charged Conducting Plates

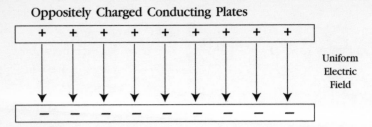

Uniform
Electric
Field

Electric Field Strength $(\vec{\in})$ is the electric force per unit of positive electric charge (at the position where the electric field strength is to be evaluated).

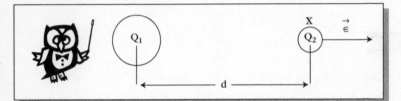

The electric field strength at **X** is

$$\vec{\in} = \frac{F_e}{Q_2}$$ ◄— magnitude of charge at **X**

$$\vec{\in} = \frac{\left[\dfrac{kQ_1Q_2}{d^2} \right]}{Q_2}$$

$$\vec{\in} = \frac{kQ_1}{d^2}$$

Unit of electric field intensity is N/C. The direction of $\vec{\in}$ is determined by first assuming Q_2 is positive and then applying the Law of Electrostatics to Q_1 and Q_2.

When two point (spherical) charges Q_1 and Q_2 are brought close to one another, the charges begin to attract or repel one another. Potential energy is built up in the electric field between the charges. (This is similar to the potential energy created when a spring is compressed or extended.) **Electric potential energy** is given by the equation:

$$E_e = \frac{kQ_1Q_2}{d}$$

Unit of Electrical Potential Energy E_e = joules

Many students experience difficulties understanding these concepts because of confusion between the concept of **electrical potential energy** defined above and another quantity, **electric potential**. To avoid confusion, a better term for electric potential is voltage. **Voltage** is defined as the energy per unit of charge (at a distance **d** from a given charge). Voltage represents the amount of energy needed to move a unit of positive charge from a position of zero voltage (zero potential) to the location **x** where the voltage is to be determined.

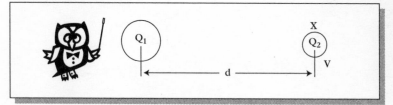

The electric potential (voltage) at **X** is

$$V = \frac{E}{Q_2} \longleftarrow \text{ magnitude of charge at } \mathbf{X}$$

$$V = \left[\frac{\dfrac{kQ_1Q_2}{d}}{Q_2}\right]$$

$$V = \frac{kQ_1}{d}$$

The unit of electric potential (voltage) is $V = J/c$ = volts.

Example:

Two small spheres are placed 40 cm apart. Sphere A is given a static electric charge of -5.0×10^{-6} C. Sphere B is given a charge of -3.0×10^{-6} C.

Find: a) the force experienced by the spheres

b) the electric field strength at B

c) the electrical potential energy in the system

d) the voltage at B

Solution:

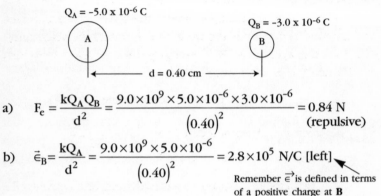

$Q_A = -5.0 \times 10^{-6}$ C

$Q_B = -3.0 \times 10^{-6}$ C

A

B

$d = 0.40$ cm

a) $F_e = \dfrac{kQ_A Q_B}{d^2} = \dfrac{9.0 \times 10^9 \times 5.0 \times 10^{-6} \times 3.0 \times 10^{-6}}{(0.40)^2} = 0.84$ N (repulsive)

b) $\vec{\epsilon}_B = \dfrac{kQ_A}{d^2} = \dfrac{9.0 \times 10^9 \times 5.0 \times 10^{-6}}{(0.40)^2} = 2.8 \times 10^5$ N/C [left]

Remember $\vec{\epsilon}$ is defined in terms of a positive charge at **B**

c) $E_e = \dfrac{kQ_A Q_B}{d} = \dfrac{9.0 \times 10^9 \times 5.0 \times 10^{-6} \times 3.0 \times 10^{-6}}{0.40} = 0.34$ J

d) $V = \dfrac{kQ_A}{d} = \dfrac{9.0 \times 10^9 \times 5.0 \times 10^{-6}}{0.40} = 1.1 \times 10^5$ V

PARALLEL PLATE CAPACITOR

A parallel plate capacitor (a pair of oppositely charged conducting plates) has a very interesting and useful electric field. Between the plates of the capacitor, the electric field is uniform. This means that a charge **Q** placed **at any position** between the plates of a capacitor will experience exactly the **same** electric force as at any other position between the plates.

Outside the plates, the field produced by each plate cancels the field produced by the other plate. There is no net electric field outside the plates of a capacitor.

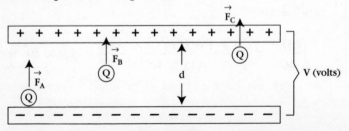

$$\boxed{\vec{F_A} = \vec{F_B} = \vec{F_C} = \text{constant}}$$

Because the force is constant between the plates of a capacitor, we can use our simple work equation **W = F d** to calculate the work done moving a charge **Q** from one plate to the other. (Remember, in situations where the force is **not** constant, the only way to determine the work done is to find the **area under the F vs. d graph** or use **W = energy gained or lost**.) Previously we learned V = E / Q. Thus, E = Q V.

But: work done = energy (gained or lost)

Between the plates of a capacitor: $\boxed{F_e\ d = Q\ V}$

Example:

In a vacuum, an electron is released from rest between the plates of a capacitor. The release point is next to the negative plate. The plates are 2.0 cm apart. If the voltage between the plates is 1000 volts, find:

    a)   the speed of the electron when it arrives at the positive plate.

    b)   the acceleration of the electron.

    c)   the electric field intensity between the plates.
$$m_e = 9.11 \times 10^{-31}\ \text{kg}$$

Solution:

a) Between the plates of a capacitor, work done = F d = QV. In a vacuum, the electron will gain only kinetic energy. (Gravitational force and gravitational potential energy are negligible compared to electrical forces and energy.)

$$\therefore \frac{1}{2}\ m_e v^2 = QV$$

$$\therefore v = \sqrt{\frac{2\ QV}{m_e}} = \sqrt{\frac{2 \times 1.6 \times 10^{-19} \times 1000}{9.11 \times 10^{-31}}} = 1.9 \times 10^7\ \text{m/s}$$

b) From the capacitor equation: $F_e d = QV$

Thus: $F_e = \dfrac{QV}{d} = \dfrac{1.6 \times 10^{-19} \times 1000}{0.020} = 8.0 \times 10^{-15}$ N

From Newton's Second Law:

$$a = \dfrac{F_e}{m_e} = \dfrac{8.0 \times 10^{-15}}{9.11 \times 10^{-31}} = 8.8 \times 10^{15} \text{ m/s}^2$$

c) Electric field strength: $\vec{\epsilon} = \dfrac{F_e}{Q}$

From the capacitor equation: $\dfrac{F_e}{Q} = \dfrac{V}{d}$

Thus: $\vec{\epsilon} = \dfrac{V}{d} = \dfrac{1000}{0.020}$

$= 5000$ N/C (toward the (–) plate)*

*Remember, $\vec{\epsilon}$ is defined in terms of **Q** being a positive charge.

PRACTICE EXERCISE

1. Two party balloons are 5.0 m apart. One balloon has a charge of $+1.6 \times 10^{-6}$ C and the other has a charge of -3.8×10^{-6} C. What is the force of attraction between the two balloons?

2. A raindrop falling through a thundercloud picks up a static electric charge of -3.6×10^{-12} C. How many extra electrons are on the raindrop?

3. A weather balloon has a charge of 2.8×10^{-5} C. What is the electric field strength 5.0 m from the balloon?

4. On a pooltable, two billiard balls are 25 cm apart. Each ball has a charge of 6.4×10^{-6} C. What is the electrical potential energy between the balls?

5. A tennis ball with a charge of 7.8×10^{-6} C is placed 2.0 m from a baseball having an electric charge of 4.2×10^{-6} C. What voltage does the tennis ball experience?

6. A capacitor in an electric power supply has plates which are 1.0 mm apart. The voltage across the plates is 120 volts. What is the electric field strength between the plates?

CHAPTER ELEVEN

Current electricity

The most useful form of electricity occurs when electric charge flows from one location to another. Moving electric charge is electric current. **Current** is defined as the rate of flow of electric charge.

Current = $\frac{charge}{time}$

$$\boxed{I = \frac{Q}{t}}$$

Q = electric charge in coulombs

t = time in seconds

I = current in C/s = amperes (A)

Electric potential (**voltage**) is the energy gained or lost per unit of charge. In physics, a good simple analogy for voltage is **pressure**. The higher the voltage (pressure pushing electrons through a circuit), the greater the current.

Voltage = $\frac{energy}{charge}$

$$\boxed{V = \frac{E}{Q}}$$

E = energy in joules

Q = charge in coulombs

V = voltage in J/C = volts

Another very useful equation can be derived by combining the above two equations:

From the first equation: Q = I t

Substituting this into the second equation: $V = \frac{E}{I\,t}$

$$\boxed{\text{Thus:} \quad E = V\,I\,t}$$

Resistance is opposition to the flow of electric current. In a current carrying wire, there are four factors that affect resistance.

1. **Length:** The longer the wire, the higher the resistance.
2. **Cross-sectional area:** The larger the cross-sectional area, the lower the resistance.
3. **Temperature:** The higher the temperature, the greater the resistance.

4. **Material:** Different materials offer different amounts of resistance to current flow. Of all metals, which are generally good **conductors**, silver is the best conductor (has the least resistance). Copper is next best, followed by aluminum. Materials that have high resistance to the flow of electric current are called **insulators** (e.g., plastic, paper, wood, glass). Between these two extremes there are materials called **semiconductors** which are moderately good at conducting electric current. Semiconductors are usually based on the elements silicon or germanium. Semiconducting materials form the basis of transistors, diodes and integrated circuits, the components of computers, pagers and cell phones.

Resistance is related to electric current and voltage by **Ohm's Law**.

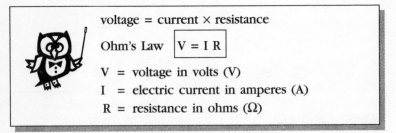

voltage = current × resistance

Ohm's Law $\boxed{V = I\,R}$

V = voltage in volts (V)
I = electric current in amperes (A)
R = resistance in ohms (Ω)

ELECTRIC CIRCUITS

Electric circuits connect by means of conductors (wires), a source of electrical energy (batteries, generators) and devices that use up electrical energy (loads or resistors). Elements in electric circuits obey Ohm's Law. The overall properties of circuits are governed by two conservation laws known as **Kirchhoff's Laws**.

Kirchhoff's Law of Voltages (conservation of energy). In an electric circuit, the sum of all the voltage increases is equal to the sum of all the voltage decreases.

Sources of electric energy (batteries, generators, etc.) increase voltage. Loads (resistances) use up electric energy and thus decrease voltage.

Kirchhoff's Law of Currents

(conservation of electric charge). At any point where wires join (a junction), the total current flowing into the junction is equal to the total current flowing out of the junction.

Kirchhoff's Laws explain the flow of electric charge and the transfer of energy in electric circuits. These two laws form the basis for all circuit analysis.

CIRCUIT SYMBOLS

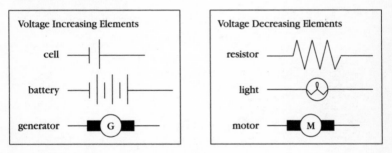

SERIES CIRCUITS

A series circuit is a circuit where there is only **one** path for the electric current to follow:

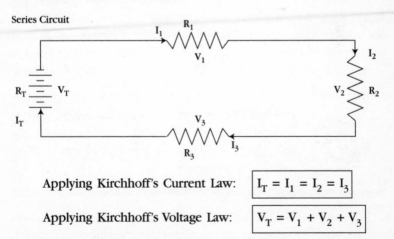

Applying Kirchhoff's Current Law: $\boxed{I_T = I_1 = I_2 = I_3}$

Applying Kirchhoff's Voltage Law: $\boxed{V_T = V_1 + V_2 + V_3}$

From Ohm's Law:

$V_1 = I_1 R_1$ Substituting into the voltage equation above:

$V_2 = I_2 R_2$ $\therefore I_T R_T = I_1 R_1 + I_2 R_2 + I_3 R_3$

$V_3 = I_3 R_3$ All the currents are equal

$V_T = I_T R_T$ $\therefore I R_T = I[R_1 + R_2 + R_3]$

$$\therefore \boxed{R_T = R_1 + R_2 + R_3}$$

Example:

Solve the series circuit shown below. (Find the current, voltage and resistance for each of the elements in the circuit.)

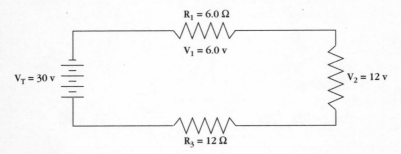

Solution:

Applying Ohm's Law to R_1: $I_1 = \frac{V_1}{R_1} = \frac{6.0}{6.0} = 1.0$ A

There is only one path for the electric current. Thus I_2, I_3 and I_T must also be 1.0.A.

Applying Ohm's Law to R_2: $R_2 = \frac{V_2}{I_2} = \frac{12}{1.0} = 12\Omega$

Applying Kirchhoff's Voltage Law: $V_T = V_1 + V_2 + V_3$
$$\therefore 30 = 6.0 + 12 + V_3$$
$$\therefore V_3 = 12 \text{ V}$$

Also $R_T = R_1 + R_2 + R_3 = 6.0 + 12 + 12 = 30\Omega$

PARALLEL CIRCUITS

A parallel circuit is a circuit where there is **more than one** path for the electric current to follow:

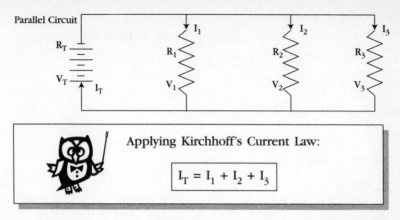

Parallel Circuit

Applying Kirchhoff's Current Law:

$$I_T = I_1 + I_2 + I_3$$

In this circuit, some electrons go from R1 to the battery and back to R1. For these electrons: $V_1 = V_T$ (Kirchhoff's Voltage Law)

Some electrons go from R_2 to the battery and back to R_2. For these electrons: $V_2 = V_T$ (Kirchhoff's Voltage Law)

Some electrons go from R_3 to the battery and back to R_3. For these electrons: $V_3 = V_T$ (Kirchhoff's Voltage Law)

Thus: $V_T = V_1 = V_2 = V_3$

Example:

For Ohm's Law:

$$I_1 = \frac{V_1}{R_1}$$

$$I_2 = \frac{V_2}{R_2}$$

$$I_3 = \frac{V_3}{R_3}$$

$$I_T = \frac{V_T}{R_T}$$

Substituting into the current equation above

$$\frac{V_T}{R_T} = \frac{V_1}{R_1} + \frac{V_2}{R_2} + \frac{V_3}{R_3}$$

All the voltages are equal

$$\therefore V_T\left[\frac{1}{R_T}\right] = V_T\left[\frac{1}{R_1} + \frac{1}{R_2} + \frac{1}{R_3}\right]$$

$$\therefore \frac{1}{R_T} = \frac{1}{R_1} + \frac{1}{R_2} + \frac{1}{R_3}$$

99

Solve the parallel circuit shown below. (Find the current, voltage and resistance for each of the elements in the circuit.)

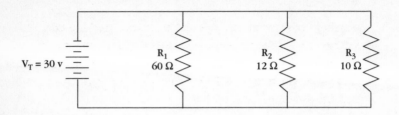

Solution:

Applying the voltage law for parallel circuits:

$$V_T = V_1 = V_2 = V_3 = 30 \text{ volts}$$

Applying Ohm's Law to each resistor:

$$I_1 = \frac{V_1}{R_1} = \frac{30}{6.0} = 5.0 \text{ A} \qquad I_2 = \frac{V_2}{R_2} = \frac{30}{12} = 2.5 \text{A}$$

$$I_3 = \frac{V_3}{R_3} = \frac{30}{10} = 3.0 \text{ A}$$

From the current equation for parallel circuits:

$$I_T = I_1 + I_2 + I_3 = 5.0 + 2.5 + 3.0 = 10.5 \text{A}$$

Using the equation for resistant in a parallel circuit:

$$\frac{1}{R_T} = \frac{1}{R_1} + \frac{1}{R_2} + \frac{1}{R_3}$$

$$\therefore \frac{1}{R_T} = \frac{1}{6.0} + \frac{1}{12} + \frac{1}{10}$$

$$\therefore \frac{1}{R_T} = \frac{1}{6.0}\left[\frac{10}{10}\right] + \frac{1}{12}\left[\frac{5}{5}\right] + \frac{1}{10}\left[\frac{6}{6}\right] = \frac{21}{60}$$

$$\therefore R_T = \frac{60}{21} = 2.86 = 2.9\Omega$$

Note that there are other possible methods of solving this circuit. For example, we could have found R_T using Ohm's Law.

Alternate solution for R_T: $R_T = \frac{V_T}{I_T} = \frac{30}{10.5} = 2.86 = 2.9\Omega$

PRACTICE EXERCISE

1.   In a typical lightning strike, 5000 C of electric charge is transferred to the ground in 0.25 seconds. Calculate the electric current in the lightning strike.

2. In a flashlight 1.2×10^{-3} C flows through the circuit. The batteries provide a voltage of 4.5 volts. How much energy is used?

3. The batteries powering a calculator contain 2100 J of chemical energy that can be converted to electrical energy. If the calculator draws a constant current of 2.5 mA at 3.0 volts, how long (in hours) will the calculator operate before the batteries need to be replaced?

4. Calculate the total resistance of each of the resistor combinations shown below:

a)

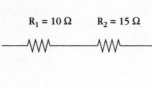

$R_1 = 10 \, \Omega$ $R_2 = 15 \, \Omega$

b)

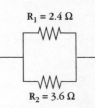

$R_1 = 2.4 \, \Omega$

$R_2 = 3.6 \, \Omega$

c)   d)

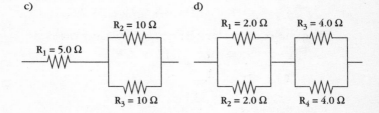

5. Solve the circuit shown below:

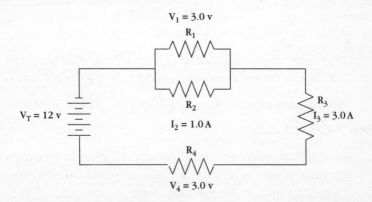

CHAPTER TWELVE

Magnetism

The ancient Greeks were aware of the magnetic properties of a certain type of iron ore called magnetite (or lodestone). The Greeks used small pieces of magnetite as compasses to navigate ships. Greek scientists also wrote of magnetite's ability to attract small bits of iron.

Today, magnets are made from many different materials. But the old terminology remains — magnets have two poles, a north pole and a south pole.

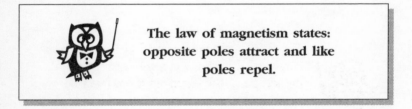

The law of magnetism states: opposite poles attract and like poles repel.

When allowed to rotate freely in the earth's magnetic field, the north pole of a magnet points toward the north magnetic pole of the earth. (Note that the "north magnetic pole" of the earth must actually be a magnetic south pole since it attracts the north poles of magnets.) The source of the earth's magnetic field lies deep within the earth and at present is not fully understood. The magnetic field of the earth interacts with charged particles streaming outward from the sun causing the Aurora Borealis (Northern Lights). The Van Allen radiation belts are regions of space surrounding the earth that contain charged particles trapped in the earth's magnetic field.

All materials display some magnetic properties. **Ferromagnetic** materials (iron, nickel and cobalt) are strongly magnetic. **Paramagnetic** materials are weakly magnetic. **Diamagnetic** materials are weakly repelled by magnetic fields.

A magnetic field is the volume of space around a magnet in which the magnet can be detected. Magnetic fields are composed of magnetic field lines. Magnetic field lines can be seen by sprinkling iron filings near a magnet. Magnetic field lines are viewed as being directed outward from north poles and inward toward south poles.

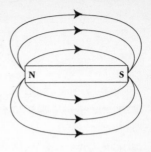

ELECTROMAGNETISM

At the beginning of the 19th century, the Danish scientist Hans Christian Oersted made a wonderful discovery. Whenever an electric charge moves through a wire, a magnetic field is created. It is on this simple principle that motors, generators and our entire electrically dependent society is based.

The direction of the magnetic field lines surrounding a current-carrying wire is given by:

The **left-hand rule for conductors** — If a conductor carrying a current is grasped with the left hand, the extended thumb will be pointing in the direction of electron flow, the curled fingers will indicate the direction of the magnetic field lines.

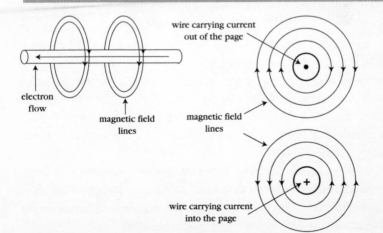

electron flow

magnetic field lines

wire carrying current out of the page

magnetic field lines

magnetic field lines

wire carrying current into the page

If we apply the **left-hand rule for conductors** to a loop of wire (see diagram), we see that magnetic field lines enter one side of the loop and exit the other. The magnetic field looks just like the magnetic field around a bar magnet. One side of the loop acts like a north pole. The opposite side of the loop acts like a south pole.

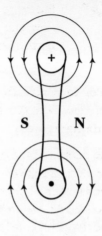

Cross-section view of a current-carrying wire loop.

When we put a series of loops together, a coil (solenoid) is created:

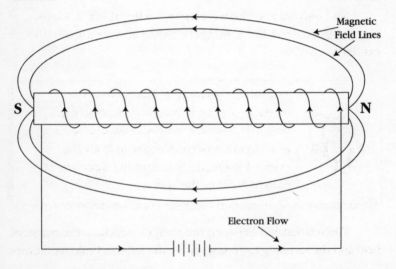

Magnetic Field Lines

Electron Flow

The left hand rule for coils helps us keep track of the relationship between the direction of electron flow, and the magnetic field lines.

The left-hand rule for coils states: Grasp the coil with the left hand using the curled fingers to indicate the direction of electron flow. The extended thumb will point at the north pole of the coil.

A coil carrying current wrapped around a ferromagnetic core forms an electromagnet. The factors which affect the strength of an electromagnet are:

1.   The current in the coil. (The greater the current, the stronger the electromagnet.)
2.   The number of turns (loops) in the coil. (The more turns, the stronger the electromagnet.)
3.   The type of core material. Some materials used as cores produce strong magnetic fields. These materials are said to have high magnetic permeability.

MOTORS

Not only can a current-carrying wire affect a magnet, as Oersted discovered, but a magnetic field can cause a force to be exerted on a wire.

The motor principle states: If a wire carrying current is placed in an external magnetic field, the wire will experience a force in a direction perpendicular to both the external magnetic field and the direction of current flow in the wire.

The relationship between the current direction, the magnetic field and the resulting force is given by the left-hand rule for motors.

The left-hand rule for motors – Use the extended fingers of the left hand to point in the direction of the lines of force (north to south pole) in the external magnetic field. Point the extended thumb of the left hand in the direction of electron flow in the wire. The palm of the left hand will face in the direction of the resulting force.

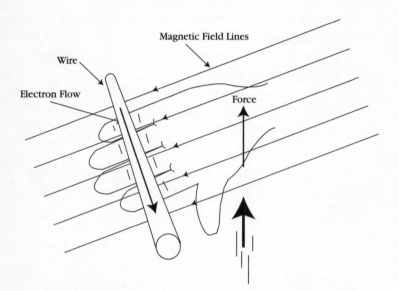

Magnetic Field Lines

Wire

Electron Flow

Force

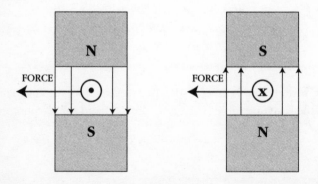

N

S

FORCE

S

N

FORCE

The motor principle, combined with a sliding contact, has been developed into an electric motor.

Direct current electric motor

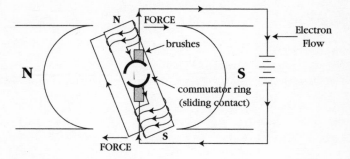

GENERATORS

In 1831, Michael Faraday discovered that magnetic fields could be used to produce an electric current. Faraday's discovery is known as:

The **principle of electromagnetic induction** - Whenever a wire moves in a magnetic field, an electric current will be caused to move in the wire. (Note that a current will also be created if the wire is at rest and the magnetic field moves or changes.)

The creation of an electric current by a magnetic field is called **induction**. The direction of the induced current is given by:

Lenz's Law
The direction of the induced current is such that the magnetic field created by the induced current will oppose the inducing magnetic field.

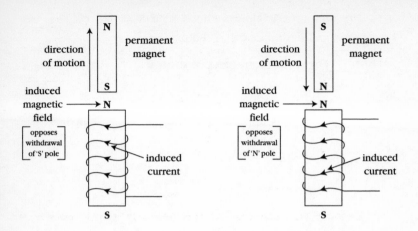

By use of sliding contacts and a rotating coil (armature), an alternating current generator can be devised.

Alternating current generator

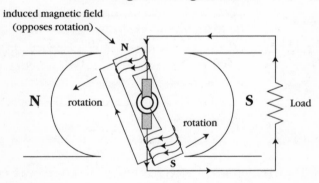

TRANSFORMERS

A transformer is a very efficient device which changes voltages in alternating current circuits. A simple transformer consists of two coils wrapped around a common iron core.

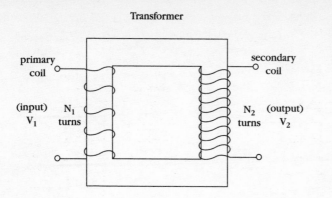

Transformer

primary coil

secondary coil

(input) V_1 N_1 turns

N_2 turns (output) V_2

Transformer Equation

$$\frac{N_1}{N_2} = \frac{V_1}{V_2}$$

Because transformers are very efficient: **input power = output power**

$$I_1V_1 = I_2V_2 \qquad \therefore P = IV$$

Example:

A transformer with an input voltage of 12 volts has 100 turns in the primary coil, and 500 turns in the secondary coil. What is the output voltage?

Solution:

$$\frac{N_1}{N_2} = \frac{V_1}{V_2} \quad \therefore \frac{100}{500} = \frac{12}{V_2} \quad \therefore V_2 = \frac{500 \times 12}{100} = 60 \text{ volts}$$

PRACTICE EXERCISE

1. Sketch the magnetic field around each of the wires shown below:

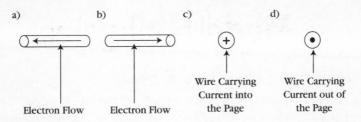

a)　　　　　b)　　　　　c)　　　　　d)

　　　　　　　　　　　　　　Wire Carrying　Wire Carrying
　　　　　　　　　　　　　　Current into　　Current out of
　　　　　　　　　　　　　　　the Page　　　　the Page

Electron Flow　　Electron Flow

2. For the current-carrying loop of wire shown in cross section below, sketch the magnetic field and indicate the location of any magnetic poles:

3. On the diagram of the direct current motor shown below indicate:

   a) the direction of the electric current in all parts of the circuit.

   b) the location of any magnetic poles.

   c) the direction of rotation of the motor armature.

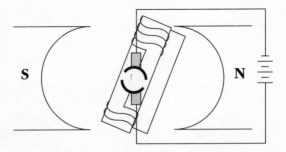

4. A transformer in a computer has a voltage of 9.0 volts across its primary coil which has 250 turns of wire. The secondary coil has an output voltage of 5.0 volts. How many turns of wire are in the secondary coil?

CHAPTER THIRTEEN

Modern physics

THE MILLIKAN EXPERIMENT

In 1906, Robert Millikan began a series of experiments that won him fame and the 1923 Nobel Prize. Millikan's experimental set-up was relatively simple. A large parallel plate capacitor was set up with the plates aligned horizontally. The top plate had a small hole drilled in it. A fine mist of oil was sprayed above the top plate. Some of the oil droplets drifted through the hole in the top plate into the region between the charged capacitor plates. An intense light source illuminated the droplets. The droplets were viewed using a microscope aimed horizontally between the plates.

This is how the Millikan experiment works: The action of creating a fine mist of oil droplets causes most of the droplets to acquire a static electric charge. Half of the charged droplets are positive, and the other half are negative. The static electric charge is created due to turbulence and friction between the oil, air and spray nozzle.

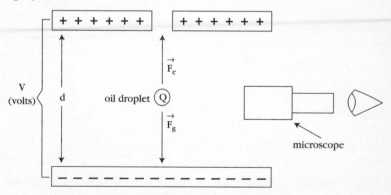

As the droplets fall under the influence of gravity, the voltage on the plates is adjusted until one droplet is captured (held motionless between the plates). A droplet that is held motionless must have an upward electric force F_e acting on it that just balances the downward gravitational force F_g. If the experiment is set up as illustrated above, with the upper plate positive, the droplet must have a negative charge in order that F_e is directed upward.

X-rays directed between the capacitor plates cause the air to ionize. By random motion, some charged air molecules come into contact with the droplet; their charge is transferred and the charge on the droplet is changed. As a result F_e changes. The droplet will begin to accelerate due to the unbalanced force. As the droplet accelerates, air friction increases to oppose the motion. Very quickly the droplet reaches a constant terminal velocity. The velocity of the droplet is measured using a stopwatch and a ruled scale in the microscope. The procedure of changing the charge and measuring the velocity of the droplet is repeated.

The results of this experiment were startling. Imagine how surprised you would feel if you surveyed a school and found that each person's birthday fell either on the 10th or 20th of the month! With hundreds of students, nobody has a birthday on the 3rd or the 12th, or the 25th. The results of the Millikan experiment were similar. The velocities of the droplet are not random. All the velocities are **integer** multiples of some basic velocity unit:

	trial	velocity (mm/s)	
	1	0	All velocities are
Simplified	2	5	multiples of 5 mm/s.
Millikan	3	–10	No other velocities
Results	4	0	were observed.
	5	–5	
	6	15	
	7	10	

Since velocities are changing by integer multiples of some basic velocity, the electric force that is causing the droplet to move must also be changing by integer multiples of some basic unit of force. But, electric force depends on the electric charge on the droplet. Thus the electric charge must be changing by integer multiples of some basic unit of charge, what we now call the elementary charge.

Millikan was the first to see the effects of individual electrons. From his experiment Millikan was able to calculate the magnitude of the elementary charge:

$$1 \text{ elementary charge} = 1.6 \times 10^{-19} \text{ coulombs}$$

Example:

An oil droplet of mass 1.35×10^{-15} kg is suspended motionless between the plates of a Millikan-type capacitor. The voltage across the plates is 550 volts and they are 4.0 cm apart. The upper plate is negative.

a) What is the charge on the droplet?

b) How many elementary charges are in excess or deficit on the droplet?

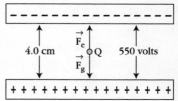

Solution:

a) This is a capacitor problem.

$$\therefore F_e d = Q V$$

Since the droplet is held motionless, the electric and gravitational forces are balanced. $\therefore F_e = F_g = mg$

Substituting in the capacitor equation above:

$$\therefore mgd = Q V$$
$$\therefore 1.35 \times 10^{-15} \times 9.8 \times 4.0 \times 10^{-2} = Q \times 550$$
$$Q = \frac{1.35 \times 10^{-15} \times 9.8 \times 4.0 \times 10^{-2}}{550}$$
$$Q = 9.6 \times 10^{-19} \text{ C}$$

b) Since the upper plate is negative and the electric force is directed upward, the charge on the droplet must be positive (a deficit of electrons). The number of electrons in deficit is:

$$N = \frac{Q}{1 \text{ el. chg.}} = \frac{9.6 \times 10^{-19}}{1.6 \times 10^{-19}} = 6 \text{ elementary charges}$$

A NEW ENERGY UNIT (THE ELECTRON-VOLT)

Imagine a single electron accelerating in a vacuum, through a potential difference of 1.0 volt. From the voltage definition:

$$V = \frac{E}{Q}$$

the kinetic energy gained by the electron is $E = Q V$.

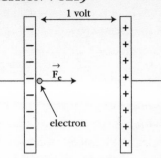

The charge on one electron is $Q = 1.6 \times 10^{-19}$ coulombs.

For a single electron: $E = Q V = 1.6 \times 10^{-19} \times 1.0 = 1.6 \times 10^{-19}$ joules

If we leave Q in terms of elementary charges rather than coulombs:

$$E = Q V = 1.0 \times 1.0 = 1.0 \text{ el. chg.} \times \text{volts}$$
$$= 1.0 \text{ electron volt (eV)}$$

But these energy calculations are for the same event. Thus the energy units must be equivalent: $\boxed{\therefore\ 1.0 \text{ eV} = 1.6 \times 10^{-19} \text{ joules}}$

THE PHOTOELECTRIC EFFECT

A negatively charged metal plate, if properly insulated, will retain its negative charge indefinitely in a dark environment. If the plate is exposed to light, however, the plate loses its negative charge. The electrons in the negatively charged plate must be gaining energy from the light and escaping from the metal. This is called the **photoelectric effect**.

Physicists who believed in the wave model of light predicted that in very dim light, it should take some time for the electrons in the metal to gain enough energy to escape. These physicists also predicted that the energy of the emitted photoelectrons should depend on the intensity of the incident light.

When these ideas were tested in the laboratory, there were a few surprises. For some colors of light, there is no time delay even in extremely dim light. The instant even dim light strikes the metal, photoelectrons are emitted. It takes a significant amount of energy

114

to knock an electron out of a metal surface. The fact that very dim light could do this, to a classical physicist, was as surprising as if a low-intensity sound (a whisper) could cause a large building to shake apart.

Another surprise was that light intensity has no effect on the energy of the emitted photoelectrons. Bright light causes more electrons to be emitted but it does not change the energy of the photoelectrons. A third unexpected observation was that some colors of light, no matter how intense, cannot produce any photoelectrons. For other colors, even very dim light easily produces large photoelectric currents.

Einstein was able to explain the photoelectric effect using a revolutionary concept. Light energy, said Einstein, is not a smooth continuous flow of energy. Light energy is packaged (quantized) in discrete energy bundles (particles). The bundle (quantum) of light energy (a photon) is the smallest unit of energy that light can have.

The quantum of light energy is:

$$E = h \, \nu$$

E = energy of photon (J or eV)
h = Planck's Constant
  = 6.6×10^{-34} Js
  = 4.1×10^{-15} eVs
ν = frequency in Hz

When a photon (a quantum) of light energy hits an electron in a metal, the photon cannot give part of its energy to the electron. The photon already has the smallest unit of energy it is possible for a photon to have. This smallest unit (quantum) cannot be further divided with part being given to the electron.

When a photon hits an electron in a metal, there are only two possible outcomes. Either the photon gives none of its energy to the electron and the photon reflects off the electron with no energy loss, or the photon gives all of its energy to the electron. If the photon gives all of its energy to the electron, the photon disappears. The electron, having gained the photon's energy, tries to pull free of the metal surface. If the electron has gained sufficient energy from the photon, it breaks free and becomes a photoelectron. The energy used up by the electron pulling free of the metal surface is called the metal **work function** (ß).

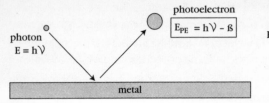

photon
$E = h\nu$

photoelectron
$\boxed{E_{PE} = h\nu - \beta}$

E_{PE} = maximum energy of the photoelectrons

metal

If we compare the photoelectric equation $E_{PE} = h\nu - \beta$ to the general equation for a straight line $y = mx + b$, we see these equations have the same mathematical form. Thus a graph of E_{PE} vs. ν will produce a straight-line graph.

On the graph, the **y** intercept is the work function –ß. The slope of the graph is Planck's Constant **h**. The threshold frequency ν_T, the **x** intercept, is the lowest frequency of light that will cause photoelectrons to be emitted.

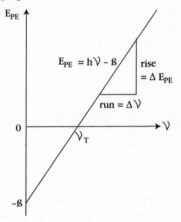

Example:

Ultraviolet light of wavelength 350 nm falls on a metal, causing photoelectrons to be emitted. The maximum energy of the photoelectrons is 2.2 eV. What is the work function of the metal?

Solution:

Using the **wave equation**: $v = \lambda\nu$

$$\therefore 3.0 \times 10^8 = 3.5 \times 10^{-7}\nu$$

$$\therefore \nu = \frac{3.0 \times 10^8}{3.5 \times 10^{-7}} = 8.57 \times 10^{14} \text{ Hz}$$

Substituting into the photoelectric equation

$$E_{PE} = h\nu - \beta$$
$$\therefore 2.2 = (4.1 \times 10^{-15})(8.57 \times 10^{14}) - \beta$$
$$\therefore 2.2 = 3.5 - \beta$$
$$\therefore \beta = 1.3 \text{ eV}$$

PARTICLE ACCELERATORS

Many particle accelerators use static electric charges to accelerate particles. The particles are accelerated in a vacuum so that none of the energy is lost due to interactions with air molecules and atoms.

Example:

An alpha particle is accelerated (in a vacuum) through a potential difference of 10 000 volts. What is the speed of the particle at the end of its acceleration? $m_a = 6.6 \times 10^{-27}$ kg

$Q_a = +2$ elementary charges

Solution:

In a vacuum, all the energy gained by the alpha particle is in the form of kinetic energy.

$$E = QV$$

$$\therefore \frac{1}{2} m v^2 = 2(1.6 \times 10^{-19}) \times 10\ 000$$

$$\therefore v = \sqrt{\frac{4(1.6 \times 10^{-19}) \times 10\ 000}{6.6 \times 10^{-27}}} = 9.8 \times 10^5 \text{ m/s}$$

THEORY OF RELATIVITY

The **Theory of Relativity** developed by Albert Einstein changed our understanding of space and time. The 1905 Theory of Relativity is based on the simple concept of an **inertial frame of reference**. An **inertial frame of reference** is any volume of space that is moving at constant velocity. Einstein's basic premise was this: For all observers in inertial frames of reference, the laws of physics are exactly the same. For observers in inertial frames of reference (for instance, the passengers on a train moving at constant velocity on perfectly smooth straight track), there is no experiment that can be done that will indicate whether the observer is moving or at rest.

A consequence of these ideas about inertial frames of reference is that the speed of light is a constant for all observers. No matter how fast we approach or withdraw from a light source, the speed of light from the source will be the same to all observers.

In the mathematics of relativity, a constant that occurs frequently is:

$$R = \sqrt{1 - \frac{v^2}{c^2}}$$

v = speed of the object
c = speed of light

Speed 'v' as a fraction of the speed of light	R
0.05	0.87
0.60	0.80
0.70	0.71
0.80	0.60
0.90	0.44
0.95	0.31
0.99	0.14

Relativity predicts that as we go faster, our clocks (time) will appear to go slower to a stationary observer. The amount of slowing is given by the equation:

$$T_v = \frac{T_s}{R}$$

T_v = time measured by observer moving at speed v
T_s = time measured by a stationary observer
R = constant described above

A second prediction of relativity is that as we speed up, the length of all objects will shrink parallel to the direction of motion by the factor:

$$L_v = L_s\, R$$

L_v = length measured by observer moving at speed v
L_s = length measured by a stationary observer
R = constant described above

Thirdly, relativity predicts that mass increases as speed increases. The mass increase is given by:

$$M_v = \frac{M_s}{R}$$

M_v = mass measured by observer moving at speed v
M_s = mass measured by a stationary observer
R = constant described above

Example:

An astronaut leaves the earth and travels in a spaceship to a distant star. The average speed of the spacecraft is 0.90 c (90% of the speed of light). The astronaut measures the time for the trip as 5.0 years. a) How long does the trip last as measured by mission control on earth? b) At rest the spacecraft is 100 m long. What length would a stationary observer measure as the ship passed at 0.90 c?

Solution:

a) $\quad T_v = \frac{T_s}{R}$

$\therefore 5.0 = \frac{T_s}{0.44}$ from chart: R = 0.44 when v = 0.90 c

$\therefore T_s = 5.0 \times 0.44 = 2.2$ years

Mission control, the stationary observer, would measure the clocks on the relativistic ship as having advanced only 2.2 years.

b) $Lv = Ls\ R = 100 \times 0.44$

$\qquad = 44\ m$

To the stationary observer, the ship would be 44 m long.

PRACTICE EXERCISE

1. In a Millikan-type experiment, a charged oil droplet of mass 1.5×10^{-14} kg is held motionless between the plates of a capacitor that are 6.0 mm apart. The capacitor voltage is 500 volts. Assuming the droplet has a negative charge, determine how many extra electrons are on the droplet.

2. In a vacuum, an iron ion (Fe^{+3}) is accelerated through a voltage of 5000 volts. How much energy is gained by the ion in:
    a) electron-volts?
    b) joules?

3. How much energy does a photon with wavelength 500 nm have in:
    a) electron-volts?
    b) joules?

4. X-rays of wavelength 170 nm strike a piece of nickel. The work function of nickel is 5.2 eV. What is the maximum energy possible for a nickel photoelectron produced by these X-rays?

5. The work function of aluminum is 4.4 eV. What is the threshold frequency for the photoelectric emission of electrons from aluminum?

6. A golf ball at rest has a mass of 250 g. Relative to a stationary observer, what is the mass of a golf ball travelling at 70% of the speed of light?

Answers to practice exercises

INTRODUCTION

1. a) 3.6×10^{-3} b) 2.5×10^{3} c) 1.2×10^{7}
   d) 2.0×10^{4} e) 1.5×10^{-3}

2. a) $50\dfrac{km}{h}\left[\dfrac{1000\ m}{1\ km}\right]\left[\dfrac{1h}{3600\ s}\right]=14\dfrac{m}{s}$

   b) $10\dfrac{m}{s^{2}}\left[\dfrac{1\ km}{1000\ m}\right]\left[\dfrac{60\ s}{1\ min}\right]^{2}=36\dfrac{km}{min^{2}}$

3. a) 310.0   b) 1012   c) 1.8
   d) 2.3   e) 1.6×10^{2}

CHAPTER 1 – MECHANICS

1. a)

b)

c)

2. a) $d = area = A_1 + A_2$ $A_1 = \frac{1}{2}(120)\ 15 = 900$

   $d = 900 + 450$ $A_2 = \frac{1}{2}(60)\ 15 = 450$

   $d = 1350\ m$

   b) $\vec{d} = A_1 - A_2$
   $= 900 - 450$
   $= 450\ m$

c) $V_{av} = \dfrac{d_{total}}{t_{total}} = \dfrac{1350}{180} = 7.5 \text{ m/s}$

d) $\vec{V}_{av} = \dfrac{\vec{d}_{total}}{t_{total}} = \dfrac{450}{180} = 2.5 \text{ m/s}$

3. $\vec{d} = \vec{v}_1 t + \dfrac{1}{2}\vec{a} t^2$

   $100 = 0 + \dfrac{1}{2}(4.0)t^2$

   $t = \sqrt{50} = 7.1 \text{ s}$

4. $\vec{v}_2^2 = \vec{v}_1^2 + 2\vec{a}\vec{d}$

   $\vec{v}_2^2 = 0^2 + 2\,(10)50$

   $\vec{v}_2 = \sqrt{1000} = 32 \text{ m/s}$

5.

$\vec{V}_{inst}$ = slope tangent = $\dfrac{rise}{run} = \dfrac{80}{5.2} = 15$ m/s

CHAPTER 2 – VECTORS

1.

$\vec{v} = 600$ m/s

a) $\vec{v}_h = v\cos60 = 600 \times 0.500 = 300$ m/s

b) $\vec{v}_v = v\sin60 = 600 \times 0.866 = 520$ m/s

2. a) $v = \dfrac{d}{t} = \dfrac{2\pi r}{t} = \dfrac{2\times\pi\times0.32}{5.0} = 0.40 \text{ m/s}$

   b) i) $\vec{v} = 0.40$ m/s (right) assuming wheel is rotating clockwise

   ii) $\vec{v} = 0.40$ m/s (down) or (up)

   iii) $\vec{v} = 0.40$ m/s (left) assuming wheel is rotating clockwise

3.

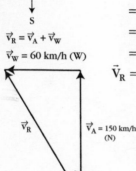

a) $V_{av} = \dfrac{\vec{d}_{total}}{t_{total}} = \dfrac{d_1 + d_2 + d_3}{t_{total}}$

$= \dfrac{2.0 + 3.0 + 6.0}{120}$

$V_{av} = 9.2 \times 10^{-2} \, km/min$

b) $\vec{V}_{av} = \dfrac{\vec{d}_{total}}{t_{total}} = \dfrac{\vec{d}_r}{t_{total}} = \dfrac{5.0}{120}$

$= 4.2 \times 10^{-2} \, km/min \, (W\,53°\,S)$

4)

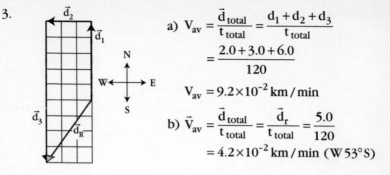

$V_R = \sqrt{V_A^2 + V_W^2}$

$= \sqrt{150^2 + 60^2}$

$= \sqrt{26\ 100}$

$= 1.6 \times 10^2$

$\vec{V}_R = 1.6 \times 10^2 \, km/h \, (N\,22°\,W)$

$\tan\theta = \dfrac{60}{150} = 0.40$

$\theta = 22°$

$\vec{v}_R = \vec{v}_A + \vec{v}_W$

$\vec{v}_W = 60 \, km/h \, (W)$

$\vec{v}_R$

$\vec{v}_A = 150 \, km/h$ (N)

θ

5. a)

$\vec{v}_R = \vec{v}_B + \vec{v}_C$

$\vec{v}_C = 1.2 \, m/s \, (east)$

$100 \, m$ $\vec{v}_B$ $2.0 \, m/s$ (north) $\vec{v}_R$

$\tan\theta = \frac{1.2}{2.0} = 0.60$

$\theta = 31°$

$V_R = \sqrt{V_B^2 + V_C^2}$

$V_R = \sqrt{2.0^2 + 1.2^2}$

$V_R = \sqrt{5.44} = 2.33$

$\vec{V}_R = 2.3 \, m/s$

$(north\,31°\,east)$

b) $t = \dfrac{d}{V_R} = \dfrac{100}{2.0} = 50 \, s$

123

CHAPTER 3 – FORCES

1. $\vec{a} = \dfrac{\vec{F}}{m} = \dfrac{80}{25} = 3.2\,\mathrm{m/s^2}$

2. $w = mg = 50 \times 9.8 = 490 = F_N$
   $F_F = \mu\, F_N = 0.46 \times 490 = 225\ \mathrm{N}$

3. Vertically:
   $v_1 = v\sin 30° = 60(0.50) = 30\,\mathrm{m/s}$
   $\vec{v}_2 = \vec{v}_1 + \vec{a}\,t$
   $-30 = 30 + (-9.8)t$
   $t = \dfrac{-60}{-9.8} = 6.12\,\mathrm{s}$

   Horizontally:
   $v_H = v\cos 30° = 60(0.87) = 52\,\mathrm{m/s}$
   $\vec{d}_H = \vec{v}_H\,t = 52 \times 6.12 = 3.2 \times 10^2\,\mathrm{m}$

4. a) $\vec{F}_R = \vec{F}_1 + \vec{F}_2$
   $F_R = \sqrt{F_1^2 + F_2^2}$
   $ = \sqrt{4000^2 + 6000^2}$
   $ = 7211$
   $\tan\theta = \dfrac{6000}{4000} = 1.5$
   $\theta = 56°$ $\vec{F}_R = 7.2 \times 10^3\ \mathrm{N\,(N\,56°\,E)}$

   b) $\vec{F}_S = 7.2 \times 10^3\ \mathrm{N\,(S\ 56°\,W)}$

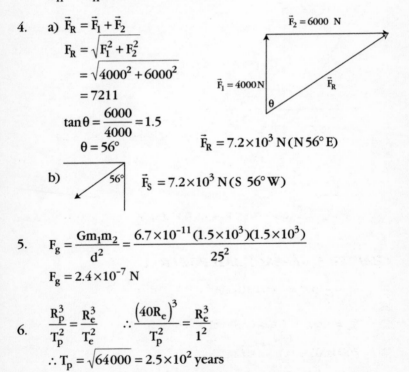

5. $F_g = \dfrac{Gm_1 m_2}{d^2} = \dfrac{6.7 \times 10^{-11}(1.5 \times 10^3)(1.5 \times 10^3)}{25^2}$
   $F_g = 2.4 \times 10^{-7}\ \mathrm{N}$

6. $\dfrac{R_p^3}{T_p^2} = \dfrac{R_e^3}{T_e^2}$ $\therefore \dfrac{(40R_e)^3}{T_p^2} = \dfrac{R_e^3}{1^2}$
   $\therefore T_p = \sqrt{64000} = 2.5 \times 10^2\ \mathrm{years}$

7. $\quad F_c = F_g \qquad R_o = R_E + 500 = 6400 + 500 = 6900 \text{km}$

$$\frac{m_s v_o^2}{R_o} = \frac{Gm_s m_E}{R_o^2} \qquad \therefore R_o = 6.9 \times 10^6 \text{m}$$

$$v_o = \sqrt{\frac{Gm_E}{R_o}} = \frac{6.7 \times 10^{-11} \times 6.0 \times 10^{24}}{6.9 \times 10^6} = 7.6 \times 10^3 \text{ m/s}$$

CHAPTER 4 – WORK

1. $\quad W = F\,d = 40 \times 6.0 = 240 \text{ J}$

2. $\quad \text{Heat} = F_f\,d = 35 \times 6.0 = 210 \text{ J}$

3.

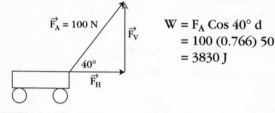

$\vec{F}_A = 100 \text{ N}$ $\quad \vec{F}_V$ $\quad 40°$ $\quad \vec{F}_H$

$$W = F_A \cos 40° \, d$$
$$= 100 \, (0.766) \, 50$$
$$= 3830 \text{ J}$$

4.

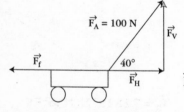

$\vec{F}_A = 100 \text{ N}$ $\quad \vec{F}_V$ $\quad \vec{F}_f$ $\quad 40°$ $\quad \vec{F}_H$

$$\vec{F}_F = -\vec{F}_H = -F_A \cos 40°$$
$$= -100 \times 0.766 = -77 \text{ N}$$

5.

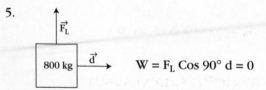

$\vec{F}_L$ $\quad$ 800 kg $\quad \vec{d}$ $\qquad W = F_L \cos 90° \, d = 0$

CHAPTER 5 – ENERGY AND POWER

1. $\quad E_k = \frac{1}{2}mv^2 = \frac{1}{2}(0.010)\,250^2 = 3.1 \times 10^2 \text{ J}$

2. $\quad E_g = mgh = 1000\,(9.8)\,60 = 5.9 \times 10^5 \text{ J}$

3. $\quad 100\,000 \text{km} = 1.0 \times 10^8 \text{m}$

$$E_g = -\frac{Gm_E m_A}{d} = -\frac{6.7 \times 10^{-11} \times 6.0 \times 10^{24} \times 2.6 \times 10^8}{1.0 \times 10^8}$$

$$= -1.0 \times 10^{15} \text{ J}$$

125

4. Initally: Finally:

$$E_T = E_g = mgh \qquad E_T' = E_g' + E_k' = mgh + \frac{1}{2}mv^2$$

$$= 25(9.8)2.0 \qquad\qquad = 20(9.8)2.0 + \frac{1}{2}(45)v^2$$

$$= 490\,J \qquad\qquad = 392 + 22.5v^2$$

By conservation of energy. $E_T = E_T'$

$$490 = 392 + 22.5v^2$$
$$v^2 = 4.35$$
$$v = 2.1\,m/s$$

5. $\Delta E = mgh = 120(9.8)\,20 = 23520\,J$

$$P = \frac{\Delta E}{t} = \frac{23520}{12} = 1960 = 2.0 \times 10^3 \text{ watts}$$

6. a) $E_{sp} = \frac{1}{2}kx^2 = \frac{1}{2}(3500)(0.75)^2 = 984 = 9.8 \times 10^2 \text{ J}$

    b) $\frac{1}{2}mv^2 = 984 \quad \therefore v = \sqrt{\frac{2.0 \times 984}{0.250}} = 89\,m/s$

CHAPTER 6 – MOMENTUM

1. a) $\vec{F}\Delta t = m\Delta\vec{v} = m(\vec{v}_2 - \vec{v}_1) = 2.0(30 - 0) = 60\,Ns$

    b) $\vec{F}\Delta t = 60 \quad \therefore \vec{F} = \frac{60}{\Delta t} = \frac{60}{0.080} = 7.5 \times 10^2 \text{ N}$

2. $\vec{p} = m\,\vec{v} = 1.2 \times 10^3\,(25) = 3.0 \times 10^4 \text{ Ns}$

3. By the Law of Conservation of Momentum
$$\vec{P}_D + \vec{P}_A = \vec{P}_{DA}$$
$$m_D\vec{v}_D + m_A\vec{v}_A = m_{DA}\vec{v}_{DA}'$$
$$0.060(12) + 0 = 0.260\vec{v}_{DA}' \quad \therefore \vec{v}_{DA}' = \frac{0.060(12)}{0.260} = 2.8\,m/s$$

4. By the Law of Conservation of Momentum
$$\vec{P}_B + \vec{P}_P = \vec{P}_B' + \vec{P}_P'$$
$$m_B\vec{v}_B + m_P\vec{v}_P = m_B\vec{v}_B' + m_P\vec{v}_P'$$
$$6.0(2.8) + 0 = 6.0\vec{v}_B' + 2.0(4.0)$$
$$16.8 = 6.0\vec{v}_B' + 8.0$$
$$\vec{v}_B' = \frac{8.8}{6.0} = 1.5\,m/s$$

5.

N, W, E, S compass diagram

$$\vec{P}_P = m_P\vec{v}_P = 1.5(6.0) = 9.0\,\text{Ns(W)}$$
$$\vec{P}_S = m_S\vec{v}_S = 0$$
$$\vec{P}_P' = m_P\vec{v}_P' = 1.5\vec{v}_P'$$
$$\vec{P}_S' = m_S\vec{v}_S' = 2.0(3.2) = 6.4\,\text{Ns}\,(\text{W}\,25°\,\text{N})$$

By the Law of Conservation of Momentum

$$\vec{P}_P + \vec{P}_S = \vec{P}_P' + \vec{P}_S'$$

$\vec{P}_S' = 6.4\,\text{Ns}$ (W 25° N)

$\vec{P}_P' \qquad 25°$

$\vec{P}_P = 9.0\,\text{Ns(W)}$

$$P_P'^2 = P_P^2 + P_S'^2 - 2P_P P_S' \cos 25$$
$$P_P'^2 = 9.0^2 + 6.4^2 - 2\,(9.0)(6.4)(9.06)$$
$$P_P'^2 = 17.55$$
$$P_P' = 4.19$$
$$1.5\vec{v}_P = 4.19$$
$$\vec{v}_P = 2.8\,\text{m/s} \ (\text{W}\,40°\,\text{S})$$

$$\frac{\sin\emptyset}{6.4} = \frac{\sin 25}{4.19}$$
$$\sin\emptyset = \frac{6.4}{4.19}\sin 25 = 0.646$$
$$\emptyset = 40°$$

CHAPTER 7 – WAVES

1. $f = \dfrac{\text{number of waves}}{\text{time}} = \dfrac{75\ 000}{60} = 1250 = 1.3 \times 10^3\,\text{Hz}$

2. $T = \dfrac{1}{f} = \dfrac{1}{512} = 2.0 \times 10^{-3}\,\text{s}$

3. a)

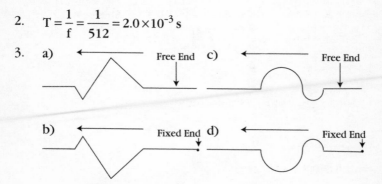

a) Free End

b) Fixed End

c) Free End

d) Fixed End

4. $v = \lambda f = 0.45 \times 740 = 3.3 \times 10^2\,\text{m/s}$

5. $f = \dfrac{c}{\lambda} = \dfrac{3.0 \times 10^8}{6.5 \times 10^{-7}} = 4.6 \times 10^{14}\,\text{Hz}$

CHAPTER 8 – LIGHT

1. $l = \dfrac{k}{d^2}$ $\qquad\qquad\qquad\qquad\qquad\qquad$ $100d_2^2 = 50(12^2)$

   $\therefore k = l_1 d_1^2 \qquad\qquad k = l_2 d_2^2 \qquad\qquad d_2^2 = 72$

   $k = 50(12^2) \qquad\qquad k = 100d_2^2 \qquad\qquad d_2 = 8.5\,\text{m}$

2. $_a n_w = \dfrac{\sin\theta_a}{\sin\theta_w}$

   $1.3 = \dfrac{\sin 60}{\sin\theta_w}$

   $\sin\theta_w = \dfrac{\sin 60}{1.3} = \dfrac{0.866}{1.3} = 0.666 \qquad \therefore \theta_w = 42°$

3. $_g n_a = \dfrac{1}{_a n_g} = \dfrac{1}{1.5} = 0.66$

4. $_d n_a = \dfrac{1}{_a n_d} = \dfrac{1}{2.4} = 0.417$

   $\sin c = {}_d n_a = 0.417 \qquad \therefore c = 25°$

5.

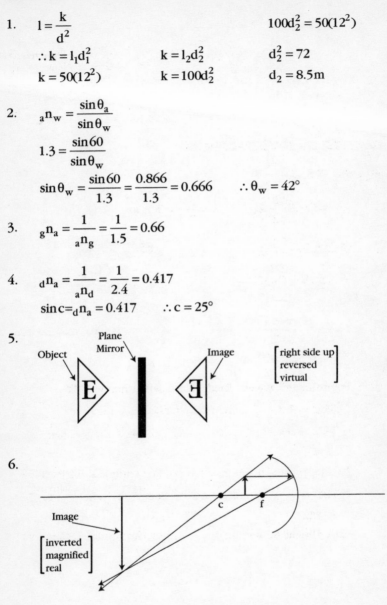

6.

7.

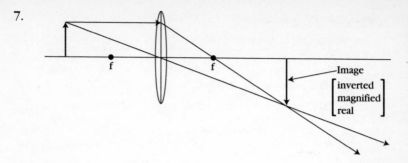

CHAPTER 9 – INTERFERENCE

1. $PS_1 - PS = (n - \frac{1}{2})\lambda$

   $23 - 18 = (8 - \frac{1}{2})\lambda \qquad \therefore \lambda = \frac{5.0}{7.5} = 0.67 \, m$

2. a) $\frac{dx_n}{L} = (n - \frac{1}{2})\lambda$

   $\frac{2.5 \times 10^{-4}(0.37)}{12} = (16 - \frac{1}{2})\lambda$

   $\therefore \lambda = \frac{2.5 \times 10^{-4}(0.37)}{15.5 \times 12}$

   $= 5.0 \times 10^{-7} \, m$

   b) $\Delta x = \frac{\lambda L}{d}$

   $\Delta x = \frac{5.0 \times 10^{-7}(12)}{2.5 \times 10^{-4}}$

   $\Delta x = 0.024 \, m = 2.4 \, cm$

3. For reflected waves: Total Phase Difference = TPD

   TPD = $2 \, t + \frac{1}{2} \, \lambda$ for each fixed end reflection

   a) TPD = $2(\lambda) + 2(\frac{1}{2} \lambda)$

   $= 3\lambda$

   Bright (Antinodal Reflection)

   b) TPD = $2(\frac{1}{2} \lambda) + 2(\frac{1}{2} \lambda)$

   $= 2\lambda$

   Bright (Antinodal Reflection)

   c) TPD = $2(\frac{3}{4} \lambda) + 2(\frac{1}{2} \lambda)$

   $= 2.5\lambda$

   Dark (Nodal Reflection)

   d) TPD = $2(\frac{1}{4} \lambda) + \frac{1}{2} \lambda$

   $= \lambda$

   Bright (Antinodal Reflection)

CHAPTER 10 – ELECTRIC CHARGE

1. $F_e = \dfrac{kQ_1Q_2}{d^2} = \dfrac{9.0\times10^9(1.6\times10^{-6})(3.8\times10^{-6})}{5.0^2} = 2.2\times10^{-3}\,N$

2. $N = \dfrac{Q}{e} = \dfrac{3.6\times10^{-12}}{1.6\times10^{-19}} = 2.25\times10^7 = 2.3\times10^7$ electrons

3. $\in = \dfrac{kQ_1}{d^2} = \dfrac{9.0\times10^9(2.8\times10^{-5})}{5.0^2} = 1.0\times10^4\,\dfrac{N}{C}$

4. $E = \dfrac{kQ_1Q_2}{d} = \dfrac{9.0\times10^9(6.4\times10^{-6})(6.4\times10^{-6})}{0.25} = 1.47 = 1.5\,J$

5. $V = \dfrac{kQ_B}{d} = \dfrac{9.0\times10^9(4.2\times10^{-6})}{2.0} = 18900 = 1.9\times10^5$ volts

6. Electric field strength: $\in = \dfrac{F_e}{Q}$

   For a parallel plate capacitor: $F_e d = QV$ $\therefore \dfrac{F_e}{Q} = \dfrac{V}{d}$

   $\therefore \in = \dfrac{V}{d} = \dfrac{120}{1.0\times10^{-3}} = 1.2\times10^5\,\dfrac{N}{C}$

CHAPTER 11 – CURRENT ELECTRICITY

1. $I = \dfrac{Q}{t} = \dfrac{5000}{0.25} = 2.0\times10^4\,A$

2. $E = QV = 1.2\times10^{-3}(4.5) = 5.4\times10^{-3}\,J$

3. $E = VIt$

   $2100 = 3.0\,(2.5\times10^{-3})\,t$

   $t = \dfrac{2100}{3.0(2.5\times10^{-3})} = 280000s = 78h$

4. a) $R_T = R_1 + R_2$

   $= 10 + 15 = 25\Omega$

   b) $\dfrac{1}{R_T} = \dfrac{1}{R_1} + \dfrac{1}{R_2}$

   $\dfrac{1}{R_T} = \dfrac{1}{2.4} + \dfrac{1}{3.6}$

   $\dfrac{1}{R_T} = 0.417 + 0.278 = 0.695$

   $R_T = \dfrac{1}{0.695} = 1.438 = 1.4\Omega$

c) $\dfrac{1}{R_X} = \dfrac{1}{R_2} + \dfrac{1}{R_3}$ d) $\dfrac{1}{R_X} = \dfrac{1}{R_1} + \dfrac{1}{R_2}$ $\dfrac{1}{R_Y} = \dfrac{1}{R_3} + \dfrac{1}{R_4}$

$\dfrac{1}{R_X} = \dfrac{1}{10} + \dfrac{1}{10}$ $\dfrac{1}{R_X} = \dfrac{1}{2} + \dfrac{1}{2}$ $\dfrac{1}{R_Y} = \dfrac{1}{4} + \dfrac{1}{4}$

$\dfrac{1}{R_X} = \dfrac{2}{10}$ $R_X = 1.0$ $R_Y = 2.0$

$R_X = 5.0$ $R_T = R_X + R_Y$

$R_T = R_1 + R_X$ $R_T = 1.0 + 2.0 = 3.0\,\Omega$

$R_T = 5.0 + 5.0 = 10\,\Omega$

5. $V_2 = V_1 = 3.0\,\text{volts}$ $V_T = V_1 + V_3 + V_4$

$R_2 = \dfrac{V_2}{I_2} = \dfrac{3.0}{1.0} = 3.0\,\Omega$ $12 = 3.0 + V_3 + 3.0$

$I_3 = I_T = 3.0\,\text{A}$ $V_3 = 6.0\,\text{volts}$

$I_1 = I_T - I_2$ $R_3 = \dfrac{V_3}{I_3} = \dfrac{6.0}{3.0} = 2.0\,\Omega$

$= 3.0 - 1.0$ $I_4 = I_3 = 3.0\,\text{A}$

$= 2.0\,\text{A}$ $R_4 = \dfrac{V_4}{I_4} = \dfrac{3.0}{3.0} = 1.0\,\Omega$

CHAPTER 12 – MAGNETISM

1.   a)   b)   c)   d)

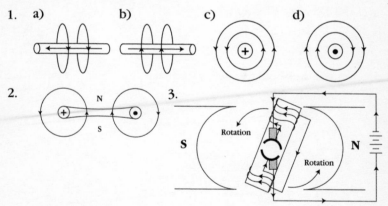

2.   3.

4. $\dfrac{N_1}{V_1} = \dfrac{N_2}{V_2}$

$\dfrac{250}{9.0} = \dfrac{N_2}{5.0}$

$N_2 = \dfrac{250\,(5.0)}{9.0} = 138.8 = 1.4 \times 10^2\ \text{turns}$

CHAPTER 13 – MODERN PHYSICS

1. $F_e d = QV$

   In equilibrium: $F_e = F_g = mg$ $N = \dfrac{Q}{e}$

   $$\therefore mgd = QV \qquad \therefore N = \dfrac{1.764 \times 10^{-18}}{1.6 \times 10^{-19}}$$

   $1.5 \times 10^{-14}(9.8)(6.0 \times 10^{-3}) = Q500$ $N = 11 \text{ electrons}$

   $$Q = \dfrac{1.5 \times 10^{-14}(9.8)(6.0 \times 10^{-3})}{500} = 1.764 \times 10^{-18} \text{ C}$$

2. a) $E = QV$ b) $E = QV$

   $\quad = 3 \,(5000)$ $\quad = 3 \,(1.6 \times 10^{-19}) \,(5000)$

   $\quad = 1.5 \times 10^4 \text{ eV}$ $\quad = 2.4 \times 10^{-15} \text{ J}$

3. $\quad \nu = \dfrac{c}{\lambda} = \dfrac{3.0 \times 10^8}{5.0 \times 10^{-7}} = 6.0 \times 10^{14} \text{ Hz}$

   a) $E = h\,\nu = 4.1 \times 10^{-15}(6.0 \times 10^{14}) = 2.5 \text{ eV}$

   b) $E = h\,\nu = 6.6 \times 10^{-34}(6.0 \times 10^{14}) = 4.0 \times 10^{-19} \text{ J}$

4. $\nu = \dfrac{c}{\lambda} = \dfrac{3.0 \times 10^8}{1.7 \times 10^{-7}} = 1.76 \times 10^{15} \text{ Hz}$

   $E = h\,\nu - ß$

   $\quad = 4.1 \times 10^{-15}(1.76 \times 10^{15}) - 5.2$

   $\quad = 2.0 \text{ eV}$

5. At the threshold frequency $\nu_T : E = 0$

   $\therefore 0 = h\,\nu_T - ß$

   $\therefore \nu_T = \dfrac{ß}{h} = \dfrac{4.4}{4.1 \times 10^{-15}} = 1.1 \times 10^{15} \text{ Hz}$

6. At 70% c: $v = 0.70c = (0.70)3.0 \times 10^8 = 2.1 \times 10^8 \text{ m/s}$

   $R = \sqrt{1 - \dfrac{v^2}{c^2}} = \sqrt{1 - \dfrac{(2.1 \times 10^8)^2}{(3.0 \times 10^8)^2}} = 0.71$

   $M_V = \dfrac{M_S}{R} = \dfrac{250}{0.71} = 3.5 \times 10^2 \text{ g}$

Possible exam questions

Uniform Motion

1. A cyclist travels at 3.0 m/s [east] for 60 s, then at 8.0 m/s [west] for 90 s, and then at 5.0 m/s [east] for 40 s.

   a) What is the **average speed** of the cyclist for the entire trip?

   b) What is the **average velocity** of the cyclist for the entire trip?

Accelerate Motion

2. A car is sitting motionless at a stoplight. The instant the light turns green, a motorcycle passes the car. The motorcycle is travelling at a constant speed of 22 m/s. At the instant the motorcycle passes, the car begins to accelerate **uniformly** at the rate of 3.5 m/s^2 .

a) How long does the car take to catch the motorcycle?

b) How far has the car travelled when the motorcycle catches up?

c) How fast is the car travelling when it catches the motorcycle?

Vectors

3. Bob can swim at 0.50 m/s in still water. He wants to swim across the Bongo Pongo river. The river is flowing **east** at 0.40 m/s. The river is 200 m wide. Bob starts from the **south** shore and he swims with his body perpendicular to the shore.

a) How long does it take Bob to cross the river?

b) What is Bob's resultant velocity while swimming?

c) What is Bob's resultant displacement?

Motion Along a Curved Path

4. A race car is observed to be initially travelling **east** at 10 m/s. Six seconds later the car, having rounded a curve, is moving N 45° W at 25 m/s. What is the **average acceleration** of the car?

Circular Motion (centripetal force)

5. A 1.00 kg stone on the end of a light string is whirled in a vertical circle of radius 1.5 m. Assume the instantaneous speed of the stone at the locations mentioned below is 5.00 m/s. Calculate the tension in the string when:

a) the string is horizontal

b) the stone is at the top of a rotation

c) the stone is at the bottom of a rotation

Forces

6. In the system below, assume the string and pulley are light and frictionless:

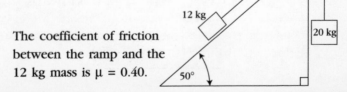

The coefficient of friction between the ramp and the 12 kg mass is μ = 0.40.

134

a) What is the force of friction acting on the 12 kg mass?

b) What is the acceleration of the system?

c) What is the tension in the string?

Gravitation

7. At what distance from the earth will a mass **m** have a force of gravity acting on it that is **half** the force of gravity that acted on **m** when it was on the earth's surface?

Orbits

8. Saturn is 40 times farther from the sun than the earth. Calculate the orbital period of Saturn in terms of **earth years**.

Momentum and Energy

9. A 5.00 kg bag of sand is hanging from the end of a long light cord. A bullet of mass 50 g, moving horizontally at 250 m/s collides with the sandbag. The bullet imbeds in the sand.

a) How high does the sandbag swing after the impact of the bullet?

b) How much heat is created by the collision?

c) What impulse does the bullet receive during the collision?

Impulse, Momentum, Energy and Springs

10. A 5.00 kg block sliding along a horizontal frictionless surface with a speed of 2.0 m/s collides with a spring as shown below:

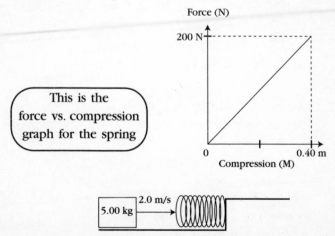

This is the force vs. compression graph for the spring

a) Is the spring shown a **linear spring**?

b) Calculate the spring constant for the spring.

c) During the collision with the spring, how much is the spring compressed when the speed of the mass is zero?

d) What impulse does the spring give to the mass during the time the mass is brought to rest?

e) How fast will the mass be moving when the spring has been compressed 15 cm? (Assume negligible heat energy is generated during the collision.)

Interference

11. a) Two radio antennas S_1 and S_2 are in phase. They emit a steady signal of 1.87×10^6 Hz. A receiver **P** placed 10.0 km from S_1 and 11.2 km from S_2 is on the 8th nodal line of the interference pattern. What is the speed of the radio waves in air as determined from this data?

b) Light is passed through two narrow slits 0.75 mm apart. A screen is placed parallel to the slits at a distance of 10.0 m. It is found that the distance on the screen from the middle of the central maximum (the line of symmetry) to the 12th nodal line is 8.45 cm. What is the wavelength of the light?

Coulomb's Law

12. Three charged spheres are arranged as shown below. **A** is **positively charged** while **B** and **C** are **negatively charged**.

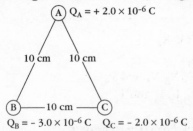

$Q_A = +2.0 \times 10^{-6}$ C

10 cm   10 cm

$Q_B = -3.0 \times 10^{-6}$ C   10 cm $Q_C = -2.0 \times 10^{-6}$ C

a) Calculate the net (resultant) electric force acting on **B**.

b) What is the electric field intensity experienced by **B**?

Work, Energy and Electric Fields

13. A small plastic sphere with an electric charge is hung from the end of a flexible insulating thread between the plates of

a capacitor as shown. The sphere hangs at an angle of 30°
to the vertical as shown.

a) Is the sphere positively or
   negatively charged?
b) If the sphere has a mass of
   3.00 g, calculate the charge on
   the sphere.
c) How many excess or missing
   electrons does this charge
   represent?
d) What is the **electric field
   intensity** experienced by the
   sphere?

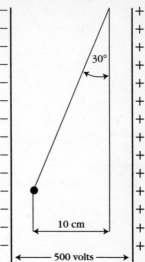

Circuits

14. Answer the following questions regarding the circuit shown
    below:

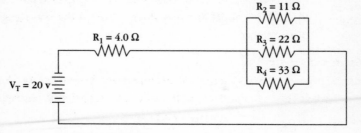

a) Solve for R_T
b) Find I_T
c) Find V_1, V_2, V_3 and V_4
d) Find I_1, I_2, I_3 and I_4

Magnetism and Electromagnetism

15. a) Sketch the magnetic field you would "see" around each
       of the following:

i) a bar magnet   ii) a wire (carrying current)

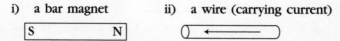

iii) a loop of wire (carrying current)

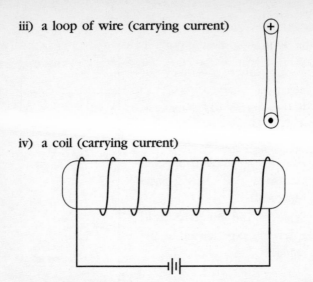

iv) a coil (carrying current)

b) On iii) and iv) above, show the location of **N** and **S** poles.
c) On what factors (three) does the strength of an electromagnet depend?
d) On the diagram below, sketch the magnetic lines of force (showing their direction) and indicate the direction of the force experienced by the wire.

N ⊙ **S**

e) Sketch the direction of current flow and the position of the **N** and **S** poles induced in the coil shown below:

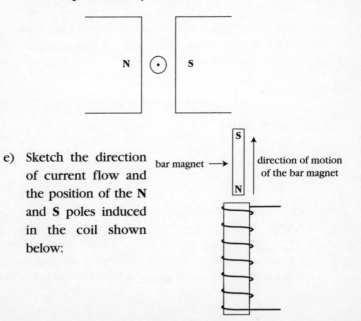

bar magnet →

direction of motion of the bar magnet

Generators

16. In the following diagram of an AC generator, show the direction of current flow and the location of the magnetic poles on the armature:

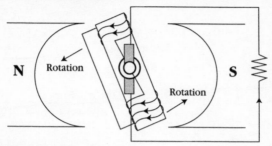

The Millikan Experiment

17. In a Millikan-type experiment, two horizontal plates are 2.0 cm apart. An oil droplet of mass 1.7×10^{-15} kg is suspended motionless between the plates when the potential difference between the plates is 520 Volts. The upper plate is **positive**.

   a) Is the droplet charged positively or negatively?
   b) Calculate the magnitude of the charge on the droplet.
   c) How many elementary charges does this charge represent?
   d) Is this charge an excess or a deficit of electrons?
   e) What is the magnitude of the electric field intensity between the plates?

Photoelectric Effect

18. Light of frequency 8.5×10^{14} Hz strikes a barium electrode. The work function of barium is 2.50 eV.

   a) Calculate the maximum energy of emitted photoelectrons.
   b) Define the term **threshold frequency**.
   c) Calculate the threshold frequency for barium.
   d) Define the term **cut-off voltage**.
   e) Calculate the cut-off voltage for this experiment.
   f) If a retarding potential of 0.50 Volts is applied to the photocell, with what maximum speed can the photoelectrons reach the plate?

Exam solutions

1. $\vec{d}_1 = \vec{v}_1 t_1 = 3.0 \times 60 = 180\,\text{m (east)}$
   $\vec{d}_2 = \vec{v}_2 t_2 = 8.0 \times 90 = 720\,\text{m (west)}$
   $\vec{d}_3 = \vec{v}_3 t_3 = 5.0 \times 40 = 200\,\text{m (east)}$

a) $V_{av} = \dfrac{d_{total}}{t_{total}} = \dfrac{d_1 + d_2 + d_3}{t_1 + t_2 + t_3} = \dfrac{180 + 720 + 200}{60 + 90 + 40} = \dfrac{1100}{190}$
   $= 5.79 = 5.8\,\text{m/s}$

b) $\vec{V}_{av} = \dfrac{\vec{d}_{total}}{t_{total}} = \dfrac{\vec{d}_1 + \vec{d}_2 + \vec{d}_3}{t_1 + t_2 + t_3} = \dfrac{180 + (-720) + 200}{60 + 90 + 40} = \dfrac{-340}{190}$
   $= -1.79$
   $\vec{V}_{av} = 1.8\,\text{m/s (west)}$

2. a)

car	motorcycle	$\vec{d}_c = \vec{d}_m$
$\vec{v}_1 = 0$	$\vec{v}_m = 22\,\text{m}$	$1.75\,t^2 = 22\,t$
$\vec{a} = 3.5\,\text{m/s}^2$	$\vec{d}_m = \vec{d}_c$	$1.75\,t^2 - 22\,t = 0$
$\vec{d}_c = \vec{d}_m$	$t_m = t_c$	$t\,(1.75\,t - 22) = 0$
$t_c = t_m$	$\vec{d}_m = \vec{v}_m\,t_m$	$t = 0$
$\vec{d}_c = \vec{v}_1\,t + \frac{1}{2}\vec{a}\,t^2$	$= 22\,t$	or
$\vec{d}_c = 0 + \frac{1}{2}\,3.5\,t^2$		$t = \frac{22}{1.75} = 12.57 = 12.6\,\text{s}$
$= 1.75\,t^2$		

b) $\vec{d}_c = \vec{d}_m = \vec{v}_m\,t_m = 22 \times 12.57 = 276.6 = 2.8 \times 10^2\,\text{m}$

c) $\vec{v}_2 = \vec{v}_1 + \vec{a}\,t = 0 + 3.5 \times 12.57 = 44\,\text{m/s}$

3. a)

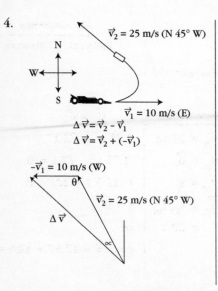

$$\vec{v}_{BP} = 0.40 \text{ m/s (E)}$$

200 m

$\vec{v}_B$
0.50 m/s
(N)

$\vec{v}_R$

θ

$$\vec{v}_R = \vec{v}_B + \vec{v}_{BP}$$

$t = \dfrac{d}{v_B}$ ⟵ must be measured in same direction

$$\therefore t = \frac{200}{0.50} = 400 \text{ s}$$

b) $V_R = \sqrt{V_B^2 + V_{BP}^2}$ $\qquad\qquad \tan\theta = \dfrac{0.40}{0.50} = 0.80$

$ = \sqrt{0.50^2 + 0.40^2}$ $\qquad\qquad \therefore \theta = 38.7$

$ = \sqrt{0.41}$ $\qquad\qquad\qquad \therefore \vec{V}_R = 0.64 \text{ m/s (N 39° E)}$

$ = 0.64 \text{ m/s}$

c) $\vec{d} = \vec{v}_R \times t = 0.64 \times 400 = 256$ m (N 39° E)

4.

$\vec{v}_2 = 25 \text{ m/s (N 45° W)}$

N

W

S

$\vec{v}_1 = 10 \text{ m/s (E)}$

$\Delta\vec{v} = \vec{v}_2 - \vec{v}_1$

$\Delta\vec{v} = \vec{v}_2 + (-\vec{v}_1)$

$-\vec{v}_1 = 10 \text{ m/s (W)}$

θ

$\vec{v}_2 = 25 \text{ m/s (N 45° W)}$

$\Delta\vec{v}$

$\propto$

$\Delta v^2 = v_1^2 + v_2^2 - 2v_1 v_2 \cos\theta$

$ = 10^2 + 25^2$

$ - 2(10)25\cos 135°$

$\Delta v = \sqrt{1079} = 32.84 \text{ m/s}$

$\dfrac{\sin\propto}{v_1} = \dfrac{\sin\theta}{\Delta v}$

$\sin\propto = \dfrac{v_1}{\Delta v} \sin\theta$

$ = \dfrac{10}{32.84} \sin 135°$

$\sin\propto = 0.2153$

$\propto = 12°$

$\vec{a} = \dfrac{\Delta\vec{v}}{t} = \dfrac{32.84}{6.0} = 5.5 \text{ m/s}^2$

(N 57° W)

141

5.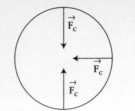

$$F_g = mg$$
$$= 1.0 \times 9.8$$
$$= 9.8\,N$$

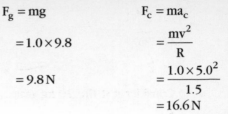

$$F_c = ma_c$$
$$= \frac{mv^2}{R}$$
$$= \frac{1.0 \times 5.0^2}{1.5}$$
$$= 16.6\,N$$

a) With the string horizontal, the **only** force acting toward the center is T, the tension in the string.

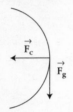

$$\vec{T} = \vec{F}_c = 17\ N$$

In this position, $\vec{F}_g$ acts perpendicular to the string and thus has no effect on the acceleration toward the center.

b) At the top:

$$\vec{F}_c = \vec{F}_g + \vec{T}$$
$$+\,16.6 = +\,9.8 + \vec{T}$$
$$\vec{T} = 6.8\ N\ (down)$$

	- up
Sign Convention	↑
	↓
	+ down

c) At the bottom:

$$\vec{F}_c = \vec{F}_g + \vec{T}$$
$$-\,16.6 = +\,9.8 + \vec{T}$$
$$\vec{T} = -26.4 = 26\ N\ (up)$$

6. a)
$$F_g = m_1 g$$
$$= 12 \times 9.8$$
$$= 117.6\ N$$

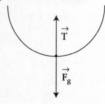

$$F_1 = F_g \mathrm{Sin}\,50°$$
$$= 117.6 \times 0.766$$
$$= 90.1\ N$$

$$F_2 = F_g' \mathrm{Cos}\,50°$$
$$= 117.6 \times 0.643$$
$$= 75.6\ N$$
$$= F_N$$
(there is no acceleration perpendicular to the ramp)

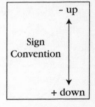

$$F_g' = m_2 g$$
$$= 20 \times 9.8$$
$$= 196\ N$$

$$F_f = \mu F_N$$
$$= 0.40 \times 75.6 = 30.2$$
$$F_f = 30\ N$$

142

b)
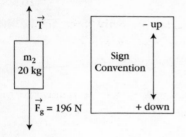

$$a = \frac{F_{un}}{m} = \frac{196 - 30 - 90}{32} = \frac{76}{32} = 2.4 \text{ m/s}^2$$

c) Consider just the 20 kg mass:

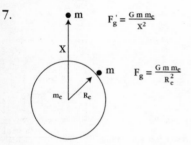

$$\vec{F}_{un} = m\vec{a}$$
$$= 20 \times 2.4$$
$$= 48 \text{ N (down)}$$
$$\vec{F}_{un} = \vec{F}_g' + \vec{T}$$
$$48 = 196 + \vec{T}$$
$$\vec{T} = 48 - 196$$
$$= -148$$
$$\vec{T} = 1.5 \times 10^2 \text{ N (up)}$$

7.

$$F_g' = \frac{G\,m\,m_e}{x^2}$$

$$F_g = \frac{G\,m\,m_e}{R_e^2}$$

$$F_g' = \frac{1}{2} F_g$$
$$\therefore \frac{G\,m\,m_e}{X^2} = \frac{1}{2} \frac{G\,m\,m_e}{R_e^2}$$
$$\therefore x^2 = 2R_e^2$$
$$\therefore x = \sqrt{2}\,R_e$$

8.   Saturn   Earth

$$K = \frac{R_s^3}{T_s^2} \qquad\qquad K = \frac{R_e^3}{T_e^2}$$

$$\therefore \frac{R_s^3}{T_s^2} = \frac{R_e^3}{T_e^2}$$
$$\therefore \frac{(40\,R_e)^3}{T_s^2} = \frac{R_e^3}{1^2}$$
$$\therefore T_s^2 = 40^3$$
$$\therefore T_s = \sqrt{64000} = 253 \text{ years}$$

9.

collision

sign convention

$-$ ← → $+$
left   right

$m_b = 0.050$ kg

$m_s = 5.0$ kg

$\vec{v}_b = 250$ m/s

This is **not** an elastic collision.
Friction acts on the bullet.

$$\vec{P}_b + \vec{P}_s = \vec{P}_{sb}\,P$$
$$m_b\vec{v}_b + m_s\vec{v}_s = m_{sb}\vec{v}_{sb}$$
$$0.050(250) + 0 = 5.050\,\vec{v}_{sb}$$
$$\vec{v}_{sb} = \frac{0.050(250)}{5.050} = 2.48 \text{ m/s}$$

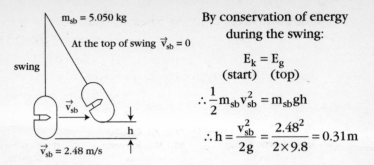

By conservation of energy during the swing:

$$E_k = E_g$$
$$\text{(start)} \quad \text{(top)}$$

$$\therefore \frac{1}{2}m_{sb}v_{sb}^2 = m_{sb}gh$$

$$\therefore h = \frac{v_{sb}^2}{2g} = \frac{2.48^2}{2 \times 9.8} = 0.31\,\text{m}$$

b) Assuming all lost E_k is converted to heat energy:

$$\text{Bullet:} \quad E_k = \frac{1}{2}m_b v_b^2 = \frac{1}{2}(0.050)(250^2) = 1563\,\text{J}$$

$$\text{Sand Bag} \quad E_k = \frac{1}{2}m_{sb}v_{sb}^2 = \frac{1}{2}(5.050)\,2.48^2 = 15.5\,\text{J}$$
$$+$$
$$\text{Bullet}$$

$$\text{Heat Energy} = \underset{\text{bullet}}{E_k} - \underset{\substack{\text{sandbag} \\ + \\ \text{bullet}}}{E_k} = 1563 - 15.5 = 1547 = 1.5 \times 10^3\,\text{J}$$

c) $\text{Impulse} = \vec{m_b}\,\Delta\,\vec{v_b} = m_b\,(\vec{v_2} - \vec{v_1}) = 0.050\,(2.48 - 250)$
$$= -247.5$$
$$\text{Impulse} = -2.5 \times 10^2\,\text{Ns}$$

10. a) Yes, this is a linear spring. The force vs. compression graph is a straight line through the point (0,0). The spring obeys Hooke's Law.

b) $k = \text{slope} = \dfrac{\text{rise}}{\text{run}} = \dfrac{200}{0.40} = 500\,\text{N/m}$

c) By the law of conservation of energy:

E_k (block before impact) $= E_{sp}$ (when spring is a maximum compression)

$$\frac{1}{2}mv^2 = \frac{1}{2}kx^2$$
$$\frac{1}{2}(5)2^2 = \frac{1}{2}(500)x^2$$
$$10 = 250x^2$$
$$x = \sqrt{0.040} = 0.20\,\text{m}$$

d) $F\,\vec{\Delta}\,t = m\,\Delta\,\vec{v} = m\,(\vec{v_2} - \vec{v_1}) = 50\,(0 - 2)$
$$= -10\,\text{Ns}$$

sign convention
$-$ ◄────► $+$
left   right

e) Potential energy of spring when compressed $= \frac{1}{2} kx^2$
$$= \frac{1}{2}(500)(0.15^2)$$
$$= 5.625 \text{ J}$$

Initial kinetic energy of block $= \frac{1}{2} mv^2$
$$= \frac{1}{2}(5) 2^2$$
$$= 10 \text{ J}$$

By the law of conservation of energy:

initial kinetic energy of block $=$ final kinetic energy of block $+$ potential energy of spring

$$10 = \frac{1}{2}mv^2 + 5.625$$

$$4.375 = \frac{1}{2}(5)v^2$$

$$v = \sqrt{\frac{2(4.375)}{5}} = 1.3 \, \text{m/s}$$

11. a) $\left|(PS_1 - PS_2)\right| = (n - \frac{1}{2})\lambda$

$$\left|(10 - 11.2)\right| = (8 - \frac{1}{2})\lambda$$

$$1.2 = 7.5\lambda$$

$$\lambda = \frac{1.2}{7.5} = 0.16 \text{km} = 160 \text{m}$$

$$v = \lambda f$$

$$= 160 \times (1.87 \times 10^6)$$

$$= 2.99 \times 10^8$$

$$v = 3.0 \times 10^8 \, \text{m/s}$$

b)

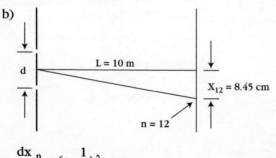

$$\frac{dx_n}{L} = (n - \frac{1}{2})\lambda$$

$$\frac{7.5 \times 10^{-4}(0.0845)}{10} = (12 - \frac{1}{2})\lambda$$

$$\lambda = \frac{7.5 \times 10^{-4}(8.45 \times 10^{-2})}{10 \times 115}$$

$$\lambda = 5.51 \times 10^{-7} \text{m}$$

12. a) $F_{BC} = \dfrac{kQ_1Q_2}{d^2} = 9.0 \times 10^9 (3.0 \times 10^{-6})(2.0 \times 10^{-6})$

   $= 5.4 \text{ N (repulsive)}$

   $F_{AB} = 5.4 \text{ N (attractive)}$

$\vec{F}_{BC} = 5.4 \text{ N}$

$\vec{F}_R = \vec{F}_{AB} + \vec{F}_{BC}$

$\vec{F}_{AB} = 5.4 \text{ N}$

   By geometry:
   $\vec{F}_R = 5.4 \text{ N} (60° \text{ to BA})$

   b) $\in = \dfrac{F_R}{Q_B} = \dfrac{5.4}{3.0 \times 10^{-6}} = 1.80 \times 10^6 \text{ N/C}$

   $\vec{\in} = 1.80 \times 10^6 \text{ N/C} (60° \text{ to BC})$

   Remember, $\vec{\in}$ is defined in terms of a + charge at B.

13. a) The sphere is positive. (It is attracted to the (–)plate.)

   b) There is equilibrium

   $\therefore \vec{F}_R = \vec{F}_e + \vec{T} + \vec{F}_g = 0$

   $\vec{T}$

   $30°$

   $\vec{F}_g = mg$
   $= 0.0030 \times 9.8$
   $= 2.94 \times 10^{-2} \text{ N}$

   $60°$

   $\vec{F}_e$

   $$\text{Tan}60° = \frac{F_g}{F_e}$$

   $$F_e = \frac{F_g}{\text{Tan}60°} = \frac{2.94 \times 10^{-2}}{1.732} = 0.01697$$

   $$F_e = 1.7 \times 10^{-2} \text{ N}$$

   This is a capacitor apparatus $\therefore F_e d = QV$

   $$1.7 \times 10^{-2}(0.10) = Q500$$

   $$Q = \frac{1.7 \times 10^{-2}(0.10)}{500} = 3.4 \times 10^{-6} \text{C}$$

   c) $N = \dfrac{Q}{e} = \dfrac{3.4 \times 10^{-6}}{1.6 \times 10^{-19}} = 2.1 \times 10^{13} \text{ elementary charges}$

d) $\in = \dfrac{F_e}{Q} = \dfrac{1.7 \times 10^{-2}}{3.4 \times 10^{-6}} = 5000 \ \text{N/C}$

14. a) Combining parallel resistors $\dfrac{1}{R_X} = \dfrac{1}{R_2} + \dfrac{1}{R_3} + \dfrac{1}{R_4}$

$\dfrac{1}{R_X} = \dfrac{1}{11} + \dfrac{1}{22} + \dfrac{1}{33}$

$\dfrac{1}{R_X} = \dfrac{6}{66} + \dfrac{3}{66} + \dfrac{2}{66} = \dfrac{11}{66}$

$R_X = 6.0\,\Omega$

$R_T = R_1 + R_X = 4.0 + 6.0 = 10\,\Omega$

b) $V_T = I_T R_T$
$20 = I_T(10)$
$I_T = 2.0\,\text{A}$

c) $V_1 = I_1 R_1$ $V_T = V_1 + V_X$
$= 2.0 \times 4.0$ $20 = 8.0 + V_X$
$= 8.0\,\text{volts}$ $V_X = 12\,\text{volts} = V_2 = V_3 = V_4$

d) $I_1 = I_T = 2.0\,\text{A}$ $I_2 = \dfrac{V_2}{R_2} = \dfrac{12}{11} = 1.09 = 1.1\,\text{A}$

$I_3 = \dfrac{V_3}{R_3} = \dfrac{12}{22} = 0.55\,\text{A}$

$I_4 = \dfrac{V_4}{R_4} = \dfrac{12}{33} = 0.36\,\text{A}$

15. a) & b)   i) 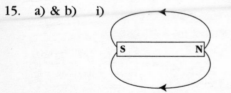   ii)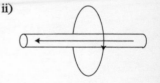

iii)    iv)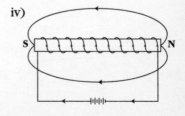

c) The strength of an electromagnet depends on:

1. the number of loops (turns) in the coil
2. the current in the coil
3. the permeability of the core material

d)

e)

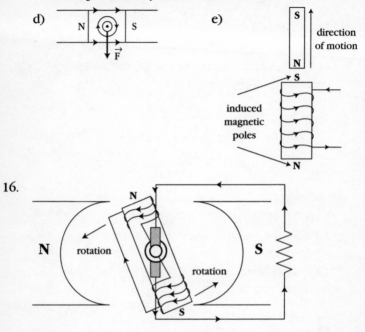

16.

17. a) The droplet is negative. It is attracted to the top (+) plate.

b)
$$F_e d = QV$$
$$mgd = QV$$
$$1.7 \times 10^{-15}(9.8)(0.020) = Q(520)$$
$$Q = \frac{1.7 \times 10^{-15}(9.8)(0.020)}{520}$$
$$= 6.41 \times 10^{-19}\,C$$

c) $N = \dfrac{Q}{e} = \dfrac{6.41 \times 10^{-19}}{1.6 \times 10^{-19}} = 4$ elementary charges

d) The droplet is negative. This is an **excess** of electrons.

e) $\vec{\epsilon} = \dfrac{F_e}{Q} = \dfrac{V}{d} = \dfrac{520}{0.020} = 26000\,N/C$ (down)

$\vec{\epsilon}$ is defined in terms
of Q being positive

18. a) $E_{PE} = h \nu - \beta$

   $= 4.1 \times 10^{-15} (8.5 \times 10^{14}) - 2.50$

   $= 3.485 - 2.50$

   $= 2.985$

   $= 3.0 \text{ eV}$

b) The threshold frequency ν_T is the minimum frequency of light that will cause photoelectrons to be emitted.

c) At the threshold frequency, the maximum energy of the emitted photoelectrons is zero.

$$E_{PE} = 0$$
$$h \nu_T - \beta = 0$$
$$4.1 \times 10^{-15} \nu_T - 2.50 = 0$$
$$\nu_T = \frac{2.50}{4.1 \times 10^{-15}} = 6.1 \times 10^{14} \text{ Hz}$$

d) The cut-off voltage Vc is the minimum voltage needed to stop the most energetic photoelectrons.

e) The most energetic photoelectrons have energy of 3.0 eV (part a).

   Electrical energy lost: $E = Q V_c$

   At the cut-off voltage: $Q V_c = 3.0$

   (1) $V_c = 3.0$ $\therefore V_c = 3.0$ volts

f) Photons are emitted with 3.0 eV of energy.

   Due to the retarding potential of 0.50 volts, the electrons lose: $E = Q V_R = (1)(0.50) = 0.50 \text{ eV}$

   Energy at plate = starting energy − energy lost due to retarding voltage.

   Energy at plate = 3.0 − 0.50 = 2.5 eV.

   Converting to joules:

$$2.5 \text{eV} \left[\frac{1.6 \times 10^{-19} \text{ joules}}{1 \text{eV}} \right] = 4.0 \times 10^{19} \text{ joules}$$

   Kinetic energy $= 4.0 \times 10^{-19}$ joules

$$\frac{1}{2} m_e v_e^2 = 4.0 \times 10^{-19} \text{ joules}$$

$$v_e = \sqrt{\frac{2 \times 4.0 \times 10^{-19}}{9.1 \times 10^{-31}}} = \sqrt{8.78 \times 10^{11}}$$

$$v_e = 9.4 \times 10^5 \text{ m/s}$$

Formulas

acceleration due to gravity $= \mathbf{g}$ $= 9.8$ m/s^2
(close to the earth's surface)

Newton's universal gravitation

constant $= \mathbf{G}$ $= 6.7 \times 10^{-11} \frac{Nm^2}{kg^2}$

radius of the earth $= \mathbf{R_e}$ $= 6.4 \times 10^6$ m

mass of the earth $= \mathbf{M_e}$ $= 6.0 \times 10^{24}$ kg

Coulomb's constant $= \mathbf{k}$ $= 2.3 \times 10^{-28} \frac{Nm^2}{el\ chg^2}$

$= 9.0 \times 10^9 \frac{Nm^2}{C^2}$

mass of the electron $= \mathbf{m_e}$ $= 9.11 \times 10^{-31}$ kg

mass of the proton $= \mathbf{m_p}$ $= 1.67 \times 10^{-27}$ kg

1 elementary charge $= \mathbf{e}$ $= 1$ el chg

$= 1.6 \times 10^{-19}$ C

speed of light in vacuum $= \mathbf{c}$ $= 3.0 \times 10^8$ m/s2

Planck's constant $= \mathbf{h}$ $= 6.6 \times 10^{-34}$ Js

$= 4.1 \times 10^{-15}$ eVs

one electron volts $= \mathbf{1\ eV} = 1.6 \times 10^{-19}$ J

Light Spectrum	
ultraviolet	< 400 nm
violet	400-460 nm
blue	460-490 nm
green	490-570 nm
yellow	570-580 nm
orange	580-600 nm
red	600-700 nm
infrared	>700 nm

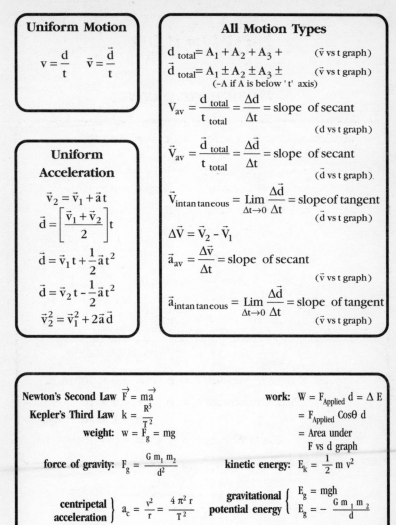

Uniform Motion

$$v = \frac{d}{t} \qquad \vec{v} = \frac{\vec{d}}{t}$$

Uniform Acceleration

$$\vec{v}_2 = \vec{v}_1 + \vec{a}\,t$$

$$\vec{d} = \left[\frac{\vec{v}_1 + \vec{v}_2}{2}\right] t$$

$$\vec{d} = \vec{v}_1 t + \frac{1}{2}\vec{a}\,t^2$$

$$\vec{d} = \vec{v}_2 t - \frac{1}{2}\vec{a}\,t^2$$

$$\vec{v}_2^2 = \vec{v}_1^2 + 2\vec{a}\,\vec{d}$$

All Motion Types

$$d_{total} = A_1 + A_2 + A_3 + \qquad (\vec{v} \text{ vs t graph})$$

$$\vec{d}_{total} = A_1 \pm A_2 \pm A_3 \pm \qquad (\vec{v} \text{ vs t graph})$$
$$(\text{-A if A is below 't' axis})$$

$$V_{av} = \frac{d_{total}}{t_{total}} = \frac{\Delta d}{\Delta t} = \text{slope of secant}$$
$$(\text{d vs t graph})$$

$$\vec{V}_{av} = \frac{\vec{d}_{total}}{t_{total}} = \frac{\Delta \vec{d}}{\Delta t} = \text{slope of secant}$$
$$(\vec{d} \text{ vs t graph})$$

$$\vec{V}_{intantaneous} = \lim_{\Delta t \to 0} \frac{\Delta \vec{d}}{\Delta t} = \text{slope of tangent}$$
$$(\vec{d} \text{ vs t graph})$$

$$\Delta \vec{V} = \vec{V}_2 - \vec{V}_1$$

$$\vec{a}_{av} = \frac{\Delta \vec{v}}{\Delta t} = \text{slope of secant}$$
$$(\vec{v} \text{ vs t graph})$$

$$\vec{a}_{intantaneous} = \lim_{\Delta t \to 0} \frac{\Delta \vec{d}}{\Delta t} = \text{slope of tangent}$$
$$(\vec{v} \text{ vs t graph})$$

Newton's Second Law $\vec{F} = m\vec{a}$

Kepler's Third Law $k = \dfrac{R^3}{T^2}$

weight: $w = F_g = mg$

force of gravity: $F_g = \dfrac{G\,m_1\,m_2}{d^2}$

centripetal acceleration $\left.\right\}$ $a_c = \dfrac{v^2}{r} = \dfrac{4\pi^2 r}{T^2}$

impulse $\vec{F}\Delta t = m\,\Delta\vec{v}$

momentum: $\vec{p} = m\vec{v}$

law of conservation of momentum $\left.\right\}$ $\vec{P}_A + \vec{P}_B = \vec{P}_A' + \vec{P}_B'$

work: $W = F_{Applied}\, d = \Delta E$
$$= F_{Applied} \cos\theta\, d$$
$$= \text{Area under } F \text{ vs d graph}$$

kinetic energy: $E_k = \dfrac{1}{2} m v^2$

gravitational potential energy $\left\{\begin{array}{l} E_g = mgh \\ E_g = -\dfrac{G\,m_1\,m_2}{d} \end{array}\right.$

power: $P = \dfrac{E}{t}$

linear springs $\left\{\begin{array}{l} F_{sp} = kx \\ E_{sp} = \dfrac{1}{2} kx^2 \end{array}\right.$

all springs: $E_{sp} = W$
$$= \text{Area under } F \text{ vs d graph}$$

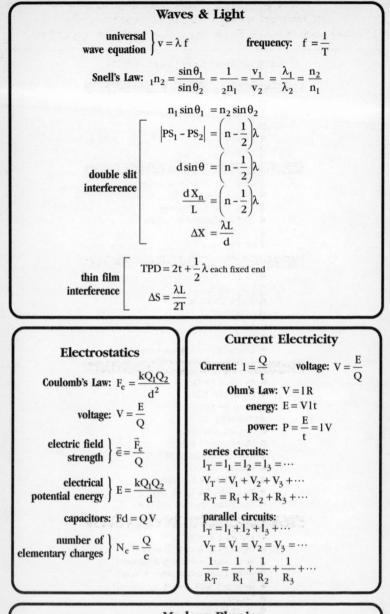

Waves & Light

universal wave equation $\Big\}$ $v = \lambda f$

frequency: $f = \dfrac{1}{T}$

Snell's Law: $_1n_2 = \dfrac{\sin\theta_1}{\sin\theta_2} = \dfrac{1}{_2n_1} = \dfrac{v_1}{v_2} = \dfrac{\lambda_1}{\lambda_2} = \dfrac{n_2}{n_1}$

$$n_1 \sin\theta_1 = n_2 \sin\theta_2$$

double slit interference $\Bigg[$

$$\left| PS_1 - PS_2 \right| = \left(n - \dfrac{1}{2} \right)\lambda$$

$$d \sin\theta = \left(n - \dfrac{1}{2} \right)\lambda$$

$$\dfrac{d X_n}{L} = \left(n - \dfrac{1}{2} \right)\lambda$$

$$\Delta X = \dfrac{\lambda L}{d}$$

thin film interference $\Bigg[$

$$TPD = 2t + \dfrac{1}{2}\lambda \text{ each fixed end}$$

$$\Delta S = \dfrac{\lambda L}{2T}$$

Electrostatics

Coulomb's Law: $F_e = \dfrac{kQ_1Q_2}{d^2}$

voltage: $V = \dfrac{E}{Q}$

electric field strength $\Big\}$ $\vec{\in} = \dfrac{\vec{F}_e}{Q}$

electrical potential energy $\Big\}$ $E = \dfrac{kQ_1Q_2}{d}$

capacitors: $Fd = QV$

number of elementary charges $\Big\}$ $N_c = \dfrac{Q}{e}$

Current Electricity

Current: $1 = \dfrac{Q}{t}$

voltage: $V = \dfrac{E}{Q}$

Ohm's Law: $V = 1R$

energy: $E = V1t$

power: $P = \dfrac{E}{t} = 1V$

series circuits:
$$1_T = 1_1 = 1_2 = 1_3 = \cdots$$
$$V_T = V_1 + V_2 + V_3 + \cdots$$
$$R_T = R_1 + R_2 + R_3 + \cdots$$

parallel circuits:
$$1_T = 1_1 + 1_2 + 1_3 + \cdots$$
$$V_T = V_1 = V_2 = V_3 = \cdots$$
$$\dfrac{1}{R_T} = \dfrac{1}{R_1} + \dfrac{1}{R_2} + \dfrac{1}{R_3} + \cdots$$

Modern Physics

photon energy: $E = h\nu$

photoelectric equation: $E = h\nu - \ss$

photon momentum: $p = \dfrac{h}{\nu}$

particle wavelength: $\lambda_p = \dfrac{h}{mv}$

energy/mass equivalence: $E = mc^2$

NOTES & UPDATES